DISASTER RECOVERY AND BUSINESS CONTINUITY

A Quick Guide for Small Organizations and Busy Executives

ABOUT THE AUTHOR

Thejendra BS is a Technical Support Manager for a software development firm in Bangalore, India. He is an Engineering graduate in Electronics from Bangalore University. He started his career as a field support engineer, maintaining and troubleshooting computers, in 1989. He also worked for two years in Saudi Arabia for a Dell Computer distributor, and has travelled extensively in the Middle East. He lives in Bangalore, India. He writes business humour, technical and serious management books. He can be contacted on thejendra@yahoo.com.

Other titles by this author from ITGP:

Practical ITIL® (2006)

Disaster Recovery and Business Continuity

A Quick Guide for Small Organizations and Busy Executives

THEJENDRA BS

IT Governance Publishing

PUBLISHER'S NOTE

First published in the United Kingdom in 2007 by IT Governance Publishing.

IT Governance Publishing
IT Governance Ltd
66 Silver Street
Ely
Cambridgeshire
CB7 4JB
United Kingdom

www.itgovernance.co.uk

© Thejendra BS 2006

ISBN 978-1-905356-14-0

PREFACE

Disaster Recovery and Business Continuity is a simple book designed to explain business recovery in an easy, self-study manner. The entire book is written in a question and answer format for easy comprehension. The book is designed to be a beginner's guide for technical and business staff in organizations wishing to know what disaster recovery and business continuity are all about.

Each chapter covers one specific area of disaster recovery and business continuity planning, and contains a set of basic and essential questions, answered in simple, jargon-free language. Every effort has been made to organize the material into an introductory, self study book. Each topic is explained in a concise manner, and very few answers extend beyond a page. Practical and real world examples with a little bit of humour are used wherever necessary.

Unless stated otherwise, the names of companies and people, mentioned in the examples in this book are fictitious. But, the names of actual companies and products mentioned are the trademarks of their respective organizations.

I would like to thank both Alan Calder for inviting me to write this book and Michael Bentley for his editorial assistance.

Thejendra BS

October 2006

CONTENTS

CHAPTER 1: Introduction to disaster recovery
and business continuity 1
CHAPTER 2: Types of disaster...39
CHAPTER 3: Data disasters...43
CHAPTER 4: Virus disasters..63
CHAPTER 5: Communication disasters...............................71
CHAPTER 6: Software disasters...77
CHAPTER 7: Data centre disasters......................................83
CHAPTER 8: IT staff disasters...89
CHAPTER 9: IT vendor disasters.......................................101
CHAPTER 10: IT project failures.......................................111
CHAPTER 11: Information security....................................123
CHAPTER 12: Disaster recovery tools...............................131
CHAPTER 13: Introduction to non-IT disasters..............135
CHAPTER 14: Disaster recovery at home.........................165
CHAPTER 15: Plenty of questions.....................................173
CHAPTER 16: How do I get started?.................................183

APPENDIX 1: Sources of further information..................217
APPENDIX 2: Useful templates and checklists...............221
APPENDIX 3: Disaster recovery training and
certification 239
APPENDIX 4: Business continuity standards..................247
APPENDIX 5: Useful stuff...251
APPENDIX 6: Disaster recovery glossary.......................249

CHAPTER 1: INTRODUCTION TO DISASTER RECOVERY AND BUSINESS CONTINUITY

'Meet success like a gentleman and disaster like a man.'

Frederick Edwin Smith (1872-1930)

The business world has changed significantly in the past few years. Organizations have undergone huge transformations over the last decade. Regardless of the industry, more and more businesses are operating on a 24x7 global basis. Competition has increased dramatically. Even small organizations with fewer than a dozen employees depend on several modern technologies and worldwide competition to remain in business. Competition is now available at the click of a mouse button. To stay in business, alive and kicking, is of paramount importance to every modern organization. Nowadays, it is not possible to run any business using the same methods and processes that were used five or ten years ago.

Secondly, the advancement and easy availability of new and useful technologies have enabled thousands of organizations worldwide to implement various technologies. Today, it is not possible to run any modern organization without the use of some computer or telecom-related technology. For example, any modern organization today will require several computers, databases, Internet access, e-mail, web-hosting, telephones, etc, for running their day-to-day operations. In addition, organizations and their customers have also become heavily dependent on technology for various needs. Many technologies cannot be fully supported in-house.

Hence, high dependence on external qualified vendors is also very critical. For example, if a vendor is not able to provide timely and efficient service for a critical IT function, organizations can get into serious trouble.

The pressure to protect businesses from all kinds of threats and risks is also increasing dramatically. An important question facing organizations today is how and who can handle predictable disasters striking the business. Nobody is immune to risks, but preventing, minimizing and avoiding disasters of all kinds have become extremely important to every organization today.

This simple book will elaborate on a variety of IT and non-IT disasters that can strike an organization at any time. This book is aimed at small organizations and IT departments wishing to get a bird's eye view of the many disaster recovery and business continuity practices around. This chapter gives short descriptions and explanations of the various terms and concepts used in disaster recovery and business continuity.

A fictitious company called RockSolid Corp will be used in many examples throughout this book. The entire book is written in a frequently asked question (FAQ) format for easy and speedy reading.

What is a disaster?

A dictionary defines a disaster as an 'occurrence causing widespread destruction and distress, or a catastrophe'. In the business environment, any event or crisis that adversely affects or disables an organization's critical business functions is a disaster. According to a number of reputable surveys and studies, hundreds of organizations worldwide go out of business every year because of the disasters that strike them. Disasters can come in all shapes and sizes, from dozens of directions. Most small businesses cannot recover from major disasters and larger organizations often struggle too.

This book will concentrate on various types of disaster that can affect businesses and how they can be prevented. This can be explained through examples:

Natural disaster example: Suppose that, due to some mishap, there was a major fire in the RockSolid computer data centre, and all the main computers containing years' of data and required business applications get burnt down. This would automatically mean that none of the RockSolid employees would be able to do any work. The entire business could come to a standstill within hours. Recovering from such a disaster would require a huge amount of effort, time and money. In addition, there could be losses in terms of reputation, losing customers, insurance and legal hassles, etc.

Technical disaster example: Instead of a fire, suppose there was a serious technical fault resulting in all computers shutting down due to some deadly virus attack or a software bug. This would also mean that none of the RockSolid employees would be able to work and business would come to a standstill. Recovering from such a disaster would also require a huge amount of effort, time and money.

Example: Sudden irrecoverable loss of data

Finance Department: 'Hello, techies. Our finance server is not working. Can you fix it immediately?'

Techie: 'Which one?'

Finance Department: 'The one that we use in our department. The black system with the green keyboard.'

Techie: 'I had a look at it, but the hard disk is dead. We will have to replace it. I will call the vendor and arrange for a replacement if possible.'

Finance Department: 'What about our data?'

Techie: 'Can't recover. The disk is dead, and we have not been backing up the data of that server, because nobody told us to. Besides, you did not approve purchase of a tape drive for that machine. Your previous finance manager was maintaining the system because of confidential data.'

Finance Department: 'Gasp!!! We have all our payroll, purchasing, billing, sales and other important financial data for the entire company and customers on that machine. Five years' data!'

Techie: 'Too bad. Got to go. I have to attend another call.'

Finance Department: 'Help! Call the CFO!! Call the CEO!!! Call the Army!!!!'

A situation like this can cripple a five-year-old business within an hour. And there are other types of potential disaster. Some could even be deliberate – sabotage, theft, espionage, etc. Hence it is necessary to ensure that organizations have properly-tested plans to recover and minimize all predictable and controllable disasters at all times.

In today's competitive world, it is extremely important for organizations to take precautions to minimize as many preventable disasters as possible in order to remain in business. Today, having a proper and tested disaster recovery plan is also a mandatory audit requirement in many organizations. Naturally, organizations will not be able to safeguard themselves against all types of disaster, but they can definitely safeguard their business against many common types of preventable disasters.

What is disaster recovery?

No modern organization can run its daily operations without computers, software, telecommunications, the Internet and so on. Disasters can cripple businesses within hours, and can strike anywhere, at any time. In today's highly competitive, 24x7 global business environment the leisurely time when a business could take days and weeks to resume operations is over. Modern businesses and organizations have also become heavily dependent on technology. If a critical computer system is not working, or not available, then businesses have to virtually close down. In many cases it is almost impossible to switch over to alternative manual or legacy processes for any length of time. Today's computer systems and networks are also extremely complex and complicated. In view of the complexity and inter-dependencies of various equipment, processes, people, etc, disasters can strike at any point and at any time. Today, businesses must be able to resume operations quickly, almost to the exact point where they stopped when the disaster struck. Though awareness of disaster recovery is increasing everywhere, very few organizations are actually well-equipped to handle disasters and restore normal operations as swiftly as possible.

Disaster recovery (DR) is the methodical preparation and execution of all the steps that will be needed to speedily recover from a disaster, usually one caused by technology. Disaster recovery planning is mainly technology-focused. Technology for these purposes can consist of voice and data communications, servers and computers, databases, critical data, etc. A DR plan should have tested and proven methods to tackle and recover from all predictable and controllable IT disasters for each of the above. For example, if there is a critical server running some crucial software, then a DR plan for that system can be a standby system in an alternative location running the identical software and having daily data synchronization. In addition, the system can also have disk mirroring, tape backups, a periodic image backup, proper change management processes, etc, for added precautions.

A proper DR plan is of critical importance to any business. It should be documented and periodically updated with key staff, contact information, locations of backups, recovery procedures, communications procedures, and a testing schedule. Additional elements may be necessary depending on company size. More details are provided later in the book.

What is business continuity?

Business continuity (BC) ensures that certain business functions continue to operate in spite of disasters striking an organization. BC is a management process that identifies various risks that threaten an organization and provides measures to safeguard the interests of its key stakeholders, customers, reputation, brand value, etc. Suppose, an IT or a non-IT disaster strikes an organization. All critical staff will be busy trying to recover from the disaster. Recovering from the disaster could range from a few minutes to several days or never. But it is essential in many customer-oriented organizations to ensure that certain 'minimum' business functions 'continue' to operate even while the main disaster is being attended to. Unless the disaster is very severe and hits all areas, or is not under the control of the management, the entire organization need not come to a standstill.

BC is mainly business-focused and will concentrate on strategies and plans for various disaster events. BC planning will prepare business areas and organizations to survive serious business interruptions, and provides the ability to perform certain 'critical functions' even during a disruptive event. For example, if a major disaster strikes a bank's main computer during banking hours, the bank management may speedily decide to allow customers to still deposit and withdraw a nominal amount of cash until such time as the main computer is fixed in the background. This is business continuity, and will ensure that customers have some minimal acceptable service in spite of a disaster. Having business continuity will also help preserve the company's reputation, image, and so on.

Note: A business continuity method need not always be a technical solution. Business continuity is all about providing workable alternatives to minimize impact.

What is crisis management?

Depending on the nature of a disaster, it may be necessary to convene a group of senior managers to control adverse media reports, customer satisfaction, deserting customers, etc. This is crisis management. Crisis management is also panic prevention. For example, in the event of a major disaster in a reputable organization, suppose there was no crisis management team. Then, there could be a possibility of a newspaper publishing a negative report causing adverse impacts on the business, stock price, reputation, etc. Hence a crisis management function becomes important. A crisis management team can ensure that such situations and possibilities are controlled by proactively taking measures to minimize losses of various kinds, including reputation losses.

Figure 1: Summary and examples of concepts

Disaster	A reputable bank's main computer's hard disk failure on Monday morning during peak banking hours. Banking operations halted. Tellers cannot verify account balances or do any electronic transactions.
Disaster recovery (DR)	Technical staff repairing the computer by replacing the hard disk and restoring data as fast as possible. Repair and restore could take several hours or more than a day.
Business continuity (BC)	Bank management allowing all customers to withdraw up to one thousand dollars manually by filling in and signing a paper withdrawal slip. Paper information to be fed into the main computer later.
Crisis management (CM)	Senior executives of the bank assuring customers that the technical problem will not cause any financial loss or improper accounting to anyone.

Note: Although the academic definitions and meanings of DR and BC are different, both terms are used simultaneously in many questions in this book. The reason for this is that the answers and concepts hold good for both in many cases. The main objective of this book is to educate organizations and IT departments on practical and real-world ways of preventing various predictable disasters and continuing in business – it is not a theoretical textbook.

What is business continuity management (BCM)?

Business continuity management is managing risks to ensure that critical *business* functions can continue to provide acceptable levels of service even in the event of a major IT or non-IT disaster. For example, if the entire data centre that houses all the important servers gets damaged in a fire, electrical short circuit or some other sudden disaster, the BCM team should assist in recovering the company from such situations in previously-planned ways. Business continuity management prepares the organization for disaster recovery options *before*, *if* and *when* a disaster occurs.

If budgets and resources were unlimited, it would be possible to build a twin of the entire organization elsewhere. But such luxuries are rarely available, nor practical. The ultimate choice of which business continuity option to use for each type of disaster should be made in consultation with several departments and business managers. As stated before, a business continuity method need not always be a technical solution. The BCM team must be able to provide cost-effective and acceptable disaster prevention solutions. Risk management is the process of identifying risks and deciding what to do about them. Risk management and business continuity are increasingly important because organizations are more susceptible today to disruptions in service caused by problems in their IT environments.

Who are the real owners of DR, BC and CM?

This is a tricky question. The immediate response would be the person(s) supporting the IT equipment or operators handling the business functions. But this is incorrect. Actually, the true owners of DR, BC and CM are the business managers of an organization. An organization may have hired some IT staff or an external vendor to provide tech support and to baby-sit an important server. But, speaking from a business perspective, those IT staff, operators or external vendors are not really the owners of DR, BC or CM for the organization. For example, if the server blows to pieces the IT staff cannot be held responsible for the organization being unable to conduct its business. It is the business managers who should know or understand what is the potential loss in terms of financial, reputation or legal angles due to stoppage of various critical businesses and IT functions. They are responsible for ensuring provision of necessary budgets, manpower, resources, alternative methods, etc, to tackle and prevent disasters. It is the business manager who is the real owner of, and ultimately responsible for, DR, BC and CM. The various ways in which the business managers can demonstrate ownership are as follows:

- **Knowledge:** Understand what the loss is in terms of financial, reputation, regulatory or legal consequences for disasters related to their critical business functions or IT equipment.

- **Financial support:** Provide necessary budgets for comprehensive maintenance of hardware, software, telecom equipment, spares, backup devices, etc. For example, suppose the business managers do not approve the purchase of a good tape drive and the necessary software, or fail to enrol into hardware maintenance for an important server – the IT staff will not be able to do much in the event of a server crash, data loss or some other technical problem on that server.

- **Provide necessary manpower:** Business managers must ensure that departments have necessary and sufficient manpower in all areas. It is very common in organizations to skimp on manpower when it comes to support, maintenance, etc. There is a common saying called 'Hire an Einstein, but refuse his request for a blackboard', which is very prevalent in many organizations worldwide. Reduced manpower and facilities in critical areas will inevitably, directly or indirectly, affect the business. See question on staff ratio later.

- **Implement recommendations:** Business managers must listen to recommendations put forth by technical staff, support staff, etc, for implementing DR and BC environments. Establishing DR and BC is expensive business. Not every critical IT function can be worked around by a low-cost alternative. It is a common practice in many organizations to ignore or avoid IT and non-IT recommendations by giving standard excuses, like cost, even though organizations will be perfectly capable of affording it. Senior management must support the necessary costs and budgets for implementing all sensible recommendations, industry standards and work-arounds necessary for DR and BC.

- **Get involved:** Senior management, including the CEO, must get involved in all aspects of their organization's DR and BCP processes. Nowadays, having a BC or DR site for many organizations is a mandatory audit requirement.

- **Policies:** Just like other essential policies of HR, finance, etc, a DR and BC policy must be enforced for all critical systems by the senior management.

- **Sustained commitment:** DR and BC is a continuous exercise. Remember – a DR or BC site is like insurance and costs money constantly. It is not enough to show interest and invest some money on a one-off basis. Continuous commitment and expenditure are required to establish proper DR and BC facilities.

Why is business continuity important?

As mentioned earlier, organizations have become extremely dependent on technology for their day-to-day operations and servicing their customers. It is not possible for any modern organization to switch over to manual processes for any length of time during a business interruption. A business interruption is any event (sudden or anticipated) that can disrupt normal business at an organization's location. For example, it is not possible to switch back to manual typewriters, postal service, and hand-written documents, spreadsheets, etc, if the entire computer and e-mail network is down. Another important concern is that any major damage to the infrastructure can result in severe financial losses, loss of reputation, and may even result in closure of the business. Today most companies are inter-connected among themselves, and to the outside world via the Internet. Any technology-related or other major failures in the company can result in the company being cut off from the rest of the world. Some of the reasons why business continuity is important are listed below:

- Businesses have become extremely dependent on IT. So failures in IT are more likely to affect the business than other areas, and that impact is more likely to be severe.

- In a networked, workflow type of environment a failure can hamper many departments and units.

- IT environments have become extremely complex and inter-related, so the number of potential failure points is increasing day by day.

- When IT fails there is not enough time to recover at a leisurely pace, because of end-user, customer and other business pressures.

Clearly, the role of professional business continuity management is essential for any organization determined to remain in business.

What is the cost of a disaster?

A disaster will definitely have numerous costs and implications. It is not only the financial cost of the equipment or process that has failed. There can be hidden costs and problems. It can even have long-term cascading effects. Depending on the nature of the business, the various costs associated with a disaster could include:

- Business losses

- Reputation losses

- Losing customers

- Stock prices dipping or free-fall

- Employee productivity losses

- Billing losses

- Unnecessary expenditure

- Fines and penalties

- Lawsuits

- Travel and logistics expenses

- Insurance and other hassles

- Other industry-specific losses

Business costs: The anticipated loss of money that the company would have made if the systems were working, eg, if the company were doing its business via a website (eg, Amazon.com) it could lose thousands of dollars to its competition if the website were down even for a few hours.

Productivity costs: Number of employees affected multiplied by their hourly cost. For example, assume that the organization had hired ten external consultants at a rate of $100 an hour each for developing a software application installed on a particular file server. If that particular server was down for three hours during business hours then the organization would suffer a loss of $3,000. This is because that amount will still need to be paid to those consultants without any productive work.

Reputation costs: No specific formula exists to calculate reputation costs. They can range from a minor manageable scratch to a total crash of the company's stock value and image in the eyes of customers and the general public. For example, if the company purchase order system is down, causing purchase orders to be delayed beyond committed delivery dates, the company may run a risk of losing those orders to their competitor or loss of reputation due to not fulfilling orders in time, etc.

Direct costs: Costs for repair or replacement of the failed equipment, manpower costs, vendor costs, liabilities, etc.

Other costs: Other costs specific to the industry.

Depending on the disaster one or more of the above losses can ruin an organization – hence the importance of paying due attention to DR and BC practices and processes.

Each of the above should be considered in sufficient detail and probability of occurrence to ensure proper business continuity alternatives. Damage must be calculated in terms of revenue, reputation, security, employees, etc. Based on the study, a detailed business continuity plan should be prepared and implemented to ensure resumption of business processes following a disruption. Today, having a business continuity plan is a mandatory audit requirement for many organizations. Businesses will not be able to generate custom from other reputable organizations if they do not have a business continuity plan. For example, the RockSolid Corp may have to prove to its major external customers that it has adequate disaster recovery facilities and that RockSolid can provide essential services even in the event of a disaster.

What is a DR or BC site?

A 'DR site' is a disaster recovery site. A 'BC site' is a business continuity site. The terms are sometimes used interchangeably. Either way, it is usually an alternative site that can be used by the business if the primary or main site fails or becomes inaccessible. For example, assume that an organization provides critical technical support on various financial applications to a key external client. Suppose there is a major IT disaster in the organization preventing staff from providing support to their client. Then, as part of disaster recovery, certain identified support staff can immediately relocate to their DR or BC site and start providing technical support. Essential support can continue from there while the main site is being rectified. Of course, the DR or BC site must have the necessary IT infrastructure and facilities to provide the required level of support.

DR or BC sites can be any or all of the following, depending on organization size, importance, and so on:

- A small or fully-fledged alternative, workable office with essential technical set-up within the city.

- A small or fully-fledged alternative workable site with essential technical set-up outside the city or in a different state or even a different country.

- A branch office of an organization from where essential functions can be continued.

- An outsourced disaster recovery location provided by a third party service provider. Nowadays, many organizations provide generic or custom-made disaster recovery locations for other organizations for a fee.

- Certain activities can also be done from home if remote connectivity options are available.

What is a command centre?

A command centre is a facility with adequate phone lines and other basic facilities to begin recovery operations. Typically it is a temporary facility used by the senior management team to begin coordinating the recovery process and used until the alternative sites are functional.

Where should a DR or BC site be located?

Several factors need to be considered when establishing a DR or BC Site. It depends on the nature of the organization and its dependent items, eg, vendor services, telecom links, material availabilities, etc. Choice of DR or BC site should also consider political, geographical, natural, human and other risks associated with the DR site location. For example, a software development company that is heavily dependent on international telecom links cannot have its DR site located in a rural area where the telecom vendors cannot provide data and voice links. Whereas another organization, eg, a manufacturing company could probably have its DR or BC site with some essential equipment located anywhere where there is an electrical supply and transport facilities.

It makes business sense to have the DR or BC site located at an acceptable distance from the main site from a logistics perspective. If essential services have to start rapidly within hours or a business day from an alternative location, the DR/BC site should be located reasonably near the main site to avoid long travel and associated logistics problems. The time to travel to a DR/BCP location is a key factor in deciding where it can be located. The various factors to be considered include:

- Data transfer requirements between main and DR site.

- Periodicity and amount of data.

- Ease of travel between main and DR site.

- Availability of support services, eg, telecom vendors, computer vendors, spare parts, etc.

- Availability of power, water, etc. It is preferable to have the DR site powered by a different electrical power grid.

- Political and civil issues of the region. For example, it does not make sense to keep a DR site in a city or country that is always plagued by civil disturbances.

- Some organizations prefer to keep their DR site located in other countries. For example, many software development companies in India have an operating DR site in Singapore and have data synchronization between the two. Should a disaster strike the main site, a core essential team in Singapore can continue to provide customer support and keep their data intact.

Establishing and maintaining a ready-to-use DR or BC site is an expensive business. Fortunately, it may not be necessary to really switch over to the DR or BC site in years. But it is like insurance – one can never predict when it will be necessary.

How does one create a DR or BC plan?

Disaster recovery is not rocket science. In fact, it is more plain common sense to ensure that the business does not go down the drain due to factors within an organization's control. A DR plan must be created by involving several departments within an organization. It is not an individual effort, although an individual in a small organization may oversee it. On the lighter side, for best results it is best to have a person who is afraid and paranoid of anything and everything as a DR manager. An ancient Chinese proverb says, 'Only a coward can create the best defences'. Study the following examples:

Example 1

A ship's captain wanted sailors for his ship. So he called a dozen hefty-looking chaps and asked who in the group were brave and excellent swimmers. About five of them lifted their hands. To everyone's surprise, the captain selected the remaining seven as his sailors. When asked why he chose the cowards he replied,

'The chaps I selected do not know how to swim and are not very brave. So they will try the hardest to keep the ship afloat.'

Example 2

A young man applied for a job as a farm-hand. When the farmer asked for his qualifications, he said, 'I can sleep when the wind blows'. This puzzled the farmer, but he liked the young man and hired him nonetheless.

A few days later, the farmer and his wife were awakened in the night by a violent storm. They quickly began to check things out to see if all was secure. They found that the shutters of the farmhouse had been securely fastened. A good supply of logs had been set next to the fireplace. And the young man slept soundly. The farmer and his wife then inspected their property. They found that the farm tools had been placed in the storage shed, safe from the elements. The tractor had been moved into the garage. The harvest was already kept inside. There was drinking water in the kitchen. The barn was properly locked. Even the animals were calm. All was well. It was only then that the farmer understood the meaning of the young man's words, 'I can sleep when the wind blows'. Since the farmhand did his work loyally and faithfully when the skies were clear, he was prepared for the storm when it broke. And when the wind blew, he was not afraid. He could sleep in peace. And, indeed, he was sleeping in peace.

Moral of the story?

There was nothing dramatic or sensational in the young farm-hand's preparations. He just faithfully did what was needed each day. The story illustrates a principle that is often overlooked about being prepared for various events that occur in life. It is only when we are facing the weather that we wish we had taken care of certain things that needed attention much earlier.

Before creating a plan, every organization must classify its functions in terms of priorities and impacts. Business and technical managers must analyse their business together and rank it in terms of priorities and business impact. For example, organizations may classify all their business functions as Low, Medium and Top priorities with a business impact for each. Obviously, not everything done by the organization can be classified as top priority or high impact.

For example, a general classification could be:

- What business functions must be up-and-running within minutes or hours of a disaster striking? For example, an organization that depends heavily on e-mail for its business cannot afford to have its e-mail server down for hours and days. It may classify e-mail as top priority, and take all necessary steps to have alternative e-mail systems. Another organization that depends heavily on a web server may classify all its web systems as top priority.

- What business functions can be down for 24 hours? For example, an organization that depends occasionally on fax can classify its fax services as medium priority and can tolerate a day's downtime.

- What business functions can be down for more than 24 hours, more than two days, a week, etc? For example, certain software development projects and product development that is still in the design or development phase can tolerate a few days or weeks of downtime. These can be classified as low priority and can wait.

See chapter 16 for sample plans.

Can organizations create DR or BC plans by themselves?

It depends on several factors. If an organization has several experienced employees who know each and every business process in detail, how they work and their importance, it is possible to create a reasonably good DR or BC plan. Otherwise, it can hire external consultants or use some standard templates. Templates are detailed checklists prepared by various organizations that can be readily used to compare an organization's preparedness. For example, the fire department can provide a checklist or template that contains several checks for preventing fire. It is also possible to have a building inspected by the fire department to certify whether the building is safe or not. Similarly, a backup software manufacturer can provide a checklist of the important things to ensure during and after a backup of data.

Important tip: Things within an organization's control must get the necessary priority, budgets and importance. The following checklist can be used:

- What areas and business functions are *completely* within an organization's control? For example, computers, data, backups, etc, are usually within an organization's control for recovery. Any loss here can be handled by the organization by implementing various safeguards and budgets, using its own manpower and resources.

- What areas and business functions are *partially* within an organization's control? Here there could be some dependence on an external service provider. For example, an organization's telephone network provided by a telecom company. An organization cannot have its own independent telephone network and has to depend on telecom service providers. Problems and shutdowns in the telecom service provider affect the organization's business, but will not be within its control. If landlines don't work, perhaps mobile phones can be used temporarily until the telecom rectifies the fault.

- What areas and business functions are *outside* an organization's control? For example, if an office is situated on the 30th floor of a building and a fire erupts on the ground floor it can affect all offices. Or if there is a terrorist attack, the police may cordon off the entire street or building, preventing employees from reaching or leaving their workplaces. Business-owners will have no say or control in such matters other than co-operating with the government forces in spite of business losses. In such cases businesses may have to resort to insurance claims, alternative sites, delays, etc.

What about DR and BC assistance from external consultants?

Nowadays, disaster recovery consultancy itself is a big business. Hundreds of DR consultants and firms have sprung up all over the world claiming to be the best among the lot. It is also industry-specific. But it is not possible to get a single, good DR consultancy on a range of business and technical processes even though they may all claim to be experts in all areas. It is necessary to evaluate the need for inviting external consultants and then decide the way forward. A combination of internal and external expertise would be appropriate in most cases.

However, the best (but unknown), DR and BC consultants to start the process could be within your own organization.

Example

Here is a short story, which many claim to be true in a certain country. Sometime in the 1980s a very important nuclear reactor suddenly stopped working. The design experts, scientists, etc, struggled very hard to set it right, but were not successful. Finally, much to the opposition of the scientists, they decided to call an ex-mechanic who was involved in the installation of the reactor. The mechanic arrived, looked around for a few minutes, and tightened a bolt in one of the sections and the machinery started working. Later, he submitted a bill of $5,000 dollars for the repair. Aghast at such an atrocious amount for just tightening a bolt, they demanded an explanation for it. The ex-mechanic split the amount as follows and resubmitted the bill.

1. Service charge for tightening the bolt: $50

2. Knowing exactly which bolt to tighten: $4,950

3. Total: **$5,000**

As you see from this story, an experienced IT person, electrician, security guard, finance chap, etc, within your own organization may have enough knowledge of what they will need to run the show in the event of a disaster striking their area of work. Their experience should be used and a combination of internal experienced staff with some external consultants would be a good choice. Organizations must select DR consultants carefully and avoid those who only give superficial advice. However, it may not be easy to pin-point the right consultant or a single consultant for all your business needs. It is better to choose consultants based on the area of DR coverage. Credentials and references play an important role in selecting them. For example, hire a reputable or experienced IT person to recommend IT DR methods, a reputable financial consultant to provide financial DR methods, etc.

Ideally, a DR or BC consultant must be a 'nuts and bolts' person who can sit with your key staff to understand your needs and then recommend practical, real world solutions. For example, if your organization wants to have a DR facility for its financial systems, the consultant must sit with your finance team and understand how the system works, the software required, the type of equipment, data synchronization requirements, etc, and then recommend a suitable disaster recovery setup, and must also be able to demonstrate its working with a mock run.

The importance of practical experience: Sir Francis Bacon said long ago, 'Knowledge is Power'. Perhaps this can be modified for today's world as 'practical knowledge is power'. Though professional certifications are becoming very important for any job, practical and real world knowledge is of paramount importance. It is important to 'first learn the trade before experimenting with tricks of the trade'. Practical hands-on experience and implementation ability are the keys to good DR consultancy.

An Indian mythological story shows the importance of real world experience over pure academic excellence:

A highly learned scholar was once travelling in a boat. The boat also had several villagers and fishermen. Wishing to pass time the scholar picked up a conversation with the other passengers and started enquiring about their educational qualifications. When he realized that most of them were illiterates and had no good academic qualifications he started showing off his rich knowledge of the Vedas and Upanishads (Hindu sacred texts), and also started insulting them by teasing that they had wasted a large amount of their lives by not studying rich academic works. Suddenly, a violent storm broke out and the boat started leaking. Immediately the boatkeeper advised everyone to jump out and swim to the shore. Everyone jumped out, but the scholar started panicking and held on to the boat. When advised to jump, he shouted that he did not know how to swim. The boatkeeper replied that the scholar had wasted his entire life by not learning how to swim that and his pure academic excellence was not going to help him now, and jumped out of the boat to swim to safety.

A second example is:

A passenger plane's pilot suddenly developed a heart attack and collapsed in the middle of the flight. A frantic airhostess called the tower and shouted for help to assist in landing the plane safely. However, no one in the tower was qualified enough to guide her on how to land a plane safely. Suddenly somebody suggested that they could call an aviation professor from a reputable university nearby to help. The professor arrived promptly, picked up the microphone and started his advice in the following manner:

'My dear lady. Let me first begin with the principles of aerodynamics before we get into the theory of aircraft engines.'

Are there any international standards and qualifications for disaster recovery and business continuity?

Yes. More and more employers are looking at certification as a condition of employment and certification is now often seen as a qualifying pre-requisite for the hiring of consultants. Today, there are primarily two recognized professional institutions certifying the business continuity professional: the Business Continuity Institute (BCI, *www.thebci.org*), based in the UK, and DRI International (DRII, *www.dr.org*), based in Falls Church, Virginia, USA. Both are member-owned, not-for-profit organizations. Both offer certification at different levels.

The BCI has five membership grades:

- Student

- Affiliate of the Business Continuity Institute

- ABCI Associate of the Business Continuity Institute

- MBCI Member of the Business Continuity Institute

- FBCI Fellow of the Business Continuity Institute

DRII International has three membership grades:

- ABCP Associate Business Continuity Planner
- CBCP Certified Business Continuity Planner
- MBCP Master Business Continuity Planner

Visit the following website for further information:

www.rothstein.com/articles/contcert.html

For details of the various DR and BC standards, training and certification options, *see Appendices 3 and 4.*

What is meant by recovery time objective (RTO) and recovery point objective (RPO)?

These terms define acceptable loss limits. They are used quite frequently in disaster recovery discussions. Determining the recovery time objective (RTO) and the recovery point objective (RPO) will define how fast an organization needs to recover in order to survive and how much data loss can be tolerated.

RTO defines the timeframe (in hours or days) within which specific business operations must be restored. It answers the question: How long can a business afford to be down? For example, will the organization tolerate a downtime of one business day? If yes, the RTO is one business day.

RPO defines the point in time to which to recover. For example, an RPO can be stated as 'Data can be recovered as of 9 pm last night'. It answers the question: How much data can the business afford to lose? For example, suppose an organization takes daily backups of one of its critical servers during the night, and the server fails abruptly the next afternoon. The IT staff will only be able to restore data as of last night's backup, so the recovery point (RPO) will be to the previous day's end of business.

Organizations can prepare tables, like the one below, for all their critical systems and tackle them one by one.

Businesses must be able to resume operations quickly. Most would prefer to resume from the point at which operations got disrupted or stopped and to be able to preserve the last data entry or transaction. This means 100% data must be available on alternative systems at all times. The shorter the delay and less data lost the sooner the business can be back in action. But there is a heavy cost for this as such online systems could be very expensive. If you want to lose no data then expensive online backup and archival systems must be used. RTO and RPO must be decided in the specific context of your particular organization.

Figure 2: RTO and RPO for critical systems

System	RTO	RPO
Development server	One business day	Data restored as of previous business day.
Finance system	One day	Data restored as of previous business day.
Data connectivity	Four hours	N/A

What are the factors to be considered in a DR or BC plan?

Disasters can come in all flavours, internal and external, so different factors need to be considered for each critical system. The entire organization's processes and systems should be classified into broad categories and tackled one by one. The DR or BC selection process starts with an assessment of the potential risks, their probability and impact for your particular enterprise. Next comes a business impact analysis (BIA). This helps determine which applications and systems require the most protection, based on the value of the data and the business impact of downtime as well as other cost factors. Organizations can broadly classify risks, with their probability of occurrence and impact, as follows:

- **Technical risks:** This will cover all IT-related issues, eg, backups, data storage and retrieval, loss of IT equipment, communication failures, virus attacks, software problems, power failures, etc.

- **Non-technical risks:** This will cover areas like building security, theft, and access by unauthorized personnel, fire hazards, etc.

- **Political risks:** Change of government and policies, civil disturbances, terrorism.

- **Financial and legal risks:** Stock market manipulations, bankruptcy, fraud, financial irregularities, failure to comply with legal regulations or standards, etc.

- **Human risks:** Losing important staff to competitors, mass resignations, death/injury/illness of key staff, disgruntled employees, spies and industrial espionage, etc.

- **Reputation risks:** All factors that can affect an organization's image, eg, employee harassment, litigation, legal turmoil, bad publicity, etc.

- **Dependency risks:** If an organization depends on external organizations, vendors and even other countries for its business it could be at risk, eg, an organization like a restaurant can depend on the existence of a large company nearby: if that large company relocates, this restaurant can go out of business.

- **Natural risks:** Flood, earthquake, hurricane, wildfire, etc.

Figure 3: Simple risk analysis

Risk	Probability	Impact
Technical	High	High
Political	Low	High
Financial	Medium	High
Fire	High	High

Note: A DR and BC plan is an ongoing process. It can *never* be perfect or complete.

How often should the DR or BC plan be updated?

As DR and BC processes are ongoing exercises, the plan should be reviewed and updated at regular intervals. It should also be immediately updated when there are any changes to the infrastructure or business processes or when critical applications and systems are introduced or discontinued. Some of the key factors requiring an update of the DR and BC plan can be:

- Addition of a new critical project or department in the organization. Study the requirements of the new project and make appropriate DRP and BC arrangements.

- Increase or decrease in critical IT equipment.

- Decommissioning of a project. If a project is being shelved for some reason, remove the DRP and BC portions of that project.

- Increase in staff, equipment, bandwidth, branch offices, etc.

- Increase or decrease in number of critical software applications.

Basically, any change should prompt an update.

What should a BCM/DR checklist consist of?

The main function of checklists for DR or BC should be to guide in checking, handling and preventing disasters. The following checklists should always be ready to hand and updated:

- List of all key personnel: contact numbers, mobiles, pagers, home addresses, etc. Update the list periodically and ensure that all staff associated with DR and BCP keep the list with them at all times.

- List of emergency phone numbers: hospitals, fire services, police, ambulances, etc.

- List of critical vendors: phone numbers, contact lists, e-mail addresses, etc.

- List of mission-critical services: which mission-critical services must your organization resume right away? Which services can you delay resuming?

- Contingency plans for each of the critical systems and services.

- Who, what, when, where and how will you resume this business?

- List of essential documents, IT equipment, etc.

- List of customers, senior management and external agencies to be notified in the event of a major disaster.

… plus any other lists that may be needed. All these lists have to be verified and updated regularly.

What is a business impact analysis (BIA)?

This is a detailed analysis of the effect on the business if a specific set of IT or non-IT services is not available. It tries to determine the risks in terms of revenue loss, reputation loss, productivity loss, etc, if the IT infrastructure or other critical facilities are down due to a disaster. A BIA will consider the following:

- Impact of damage to premises, data centre, etc.

- Impact of damage to IT systems: servers, computers, networks, telecommunications, etc.

- Impact of damage to important data in terms of loss or corruption.

- Loss of key staff: IT support, business managers, etc.

- Impact on external and internal customers.

- Legal and reputation implications if disasters occur.
- Dependencies on external vendors, suppliers, etc.
- Impact of security threats: viruses, hackers who may steal confidential information, etc.
- Impact of damage and loss of power, air conditioners, etc, required for IT services.
- Damage due to sabotage, natural disasters, political threats, etc.
- Other industry-specific impacts.

For example, a very basic BIA can be as shown in Figure 4.

Figure 4: Simple business impact analysis

System	Probability	Impact of downtime
Company web server down	High	$5,000 in lost business per hour
Company network down	Medium	Productivity loss of $50 per hour per employee

Organizations can prepare such charts to decide which business functions require priority.

Who can invoke business continuity?

As part of their business continuity plans, organizations must first decide what qualifies as a disaster. Any routine equipment problem, maintenance downtime, short-term problems, etc, should not be termed disasters or alternative facilities invoked. The decision to brand an IT shutdown as a disaster must be taken only by senior business and IT managers. A business recovery team can also be constituted: this is a group of qualified and senior staff responsible for maintaining the business recovery procedures and for co-ordinating the recovery of business functions in an organization. For example, suppose that the entire IT infrastructure is down due to a power fault. If the fault is expected to be rectified within a couple of hours then the organization need not classify it as a crisis and start rushing employees to invoke their disaster recovery or business continuity procedures. On the other hand, if it is sure that the power failure is more severe and may take more than an acceptable time to restore, then the senior management may invoke the disaster recovery procedures.

The following types of disaster can necessitate invoking business continuity beyond the agreed RTO and RPO:

- Severe or major business impact.

- Adverse customer impact.

- High risk exposure to organization.

- Critical system down.

What are the options available for business continuity?

Technically and financially it is possible to build a twin of any organization. But not all organizations may want this, or can afford such a luxury. Business continuity is industry-specific. For example, the police, fire, ambulance services, etc, cannot afford to have their IT and other infrastructure down even for a few minutes. Other organizations like a small automobile spare parts manufacturer may be able to withstand an IT failure for a couple of days. Depending on the organization, size, industry, budgets, etc, companies can have a number of choices:

Manual: If it is possible to use manual methods.

Other offices: If an organization is decentralized and has many independent branches, then it may be possible to use the facilities in another branch until the affected branch comes online again.

Cold standby: Organizations can have an alternative site with basic IT and non-IT facilities that can be switched on during extended failures.

Warm standby: This involves re-establishing critical systems and services within a short period of time, usually achieved by having redundant equipment that can be used during disasters.

Hot standby: This will involve having an alternative site with continuous mirroring of live data and configurations. These sort of facilities are usually used by banks, the military, etc, where it is not possible to afford any downtime.

What is a DR or BC exercise?

One of the ways of testing the disaster recovery 'readiness' of an organization is to conduct frequent mock exercises of the various areas mentioned in the DR plan. This is usually done by simulating a crisis situation. Such mock exercises test the organization's ability to respond to a crisis in a planned and effective manner instead of becoming chaotic. For example, if the finance department server is a critical DR item, a mock exercise can be conducted on a weekend or after hours by invoking a mock disaster by shutting down the system, sending the finance chaps to work from the DR site and noting down all the issues, limitations, deficiencies, missed out items, etc. These exercises will give a first-hand feeling for how an organization or department can handle a real disaster if one occurs. Appropriate measures can then be taken to ensure proper disaster recovery. For example, the finance department may notice that it is not possible to operate the finance application without connecting at least one printer. Later, during the next mock run, a printer can be hooked up to test the finance application.

Is there any training available for disaster recovery, business continuity, etc?

Yes. There are many courses available. The Disaster Recovery Institute International (DRI International) provides various basic and advanced courses. Visit *www.drii.org* for more information. You can also visit Business Continuity Institute at *www.thebci.org*. Many universities have also started providing diploma and graduate courses on disaster recovery and business continuity.

For more information on disaster recovery and business continuity training and certification, *see Appendix 3.*

What are the biggest roadblocks for disaster recovery or business continuity?

Every businessman would like to have 100% disaster recovery and business continuity. However, very few organizations are actually willing to make the necessary investments in terms of people and budgets to ensure reliable disaster recovery and business continuity environments. Some of the biggest roadblocks that prevent proper disaster recovery or business continuity are:

- **Lack of sustained management commitment:** One of the primary roadblocks for disaster recovery will be lack of sustained top management commitment. For example, the top management may approve the establishment of a DR or BCP site at a time when they are particularly influenced by business and competitive pressures. But later they may not be willing to invest the necessary ongoing budgets and manpower to keep the site fully operational at all times.

- **Inadequate budgets:** Business managers unable or unwilling to invest sufficiently to establish DR and BC options. Disaster recovery options require investment in redundant equipment, spares, data synchronization equipment, software, hardware, training, insurance, alternative sites, etc.

- **Manpower:** Not willing to invest in additional technical staff needed to maintain and manage a DR site.

- **Knowledge:** Lack of knowledge about what is needed to establish a proper disaster recovery.

- **Other reasons:** Various internal factors and limitations.

In fact, in a large number of cases DR and BC plans just remain on paper or have insufficient capability to handle real disasters. But businesses really do need to invest in the necessary budgets and staff if they are to ensure that their businesses are safe from preventable disasters.

How much money is required to establish a proper disaster recovery facility?

It depends on various factors and the nature of the organization. Theoretically, it is possible to establish a twin of the entire organization if budgets were unlimited. However, such a setup is rarely possible or required. Broadly, DR costs can be classified as:

- **People costs:** The number of additional employees, contractors or vendors required and trained to handle various disasters.

- **IT costs:** How many additional computer systems, software licences, telephones, communication systems, etc, are required?

- **Maintenance and ongoing costs:** It is not enough to just establish a fully-fledged disaster recovery setup as a one-off exercise. The whole setup must be properly maintained and periodically updated with new systems, software, data updates, dry runs, etc. This all costs money.

- **Infrastructure and other costs:** This involves costs for building rents, electricity costs, air conditioning costs, security, transport, telephone costs, etc.

- **Other costs:** Various one-off or ongoing costs.

Do's and don'ts

Do's

- Identify a dedicated team within your organization to be responsible for DR and BC.

- Ensure each member knows exactly what he or she is supposed to do and not do.

- Clearly establish the scope of your DR and BC plans.

- Analyse all your business functions and sort them by their importance.

- Have an organizational policy to enforce DR and BC practices.

- Have periodic meetings on various DR issues and keep the DR plan updated regularly.

- Keep your customers and employees informed.

- Conduct DR exercises regularly. Keep yourself updated with new industry practices, standards and qualifications.

Don'ts

- Take DR and BC functions lightly.

- Provide give DR inadequate budgets and resources.

- Ignore internal talent and knowledge within your organization.

CHAPTER 2: TYPES OF DISASTER

'You should treat all disasters as if they were trivialities but never treat a triviality as if it were a disaster.' *Quentin Crisp*

What are the main types of risk and disaster that can strike an organization?

A general dictionary defines 'risk' as 'the possibility of suffering harm or loss or danger'. Risks and disasters come in all sizes, shapes, colours and flavours. However, for the sake of convenience, risk can be broadly classified as:

- Technical risks: disasters relating to IT, computers, networking, software, etc.

- Non-technical risks: disasters relating to other areas.

What is a technical risk?

Any organization today will use one or more of the following IT systems:

- Computers of various sizes and capacities ranging from small laptops to large mainframes.

- Data backup systems to store and retrieve large amounts of data.

- E-mail systems for internal and external communication.

- Telecommunication systems, eg, fax, dial-up lines, leased lines for connecting their offices, branches, etc, within and between cities, states and countries.

- Various software programmes, eg, office suites, databases, remote connectivity tools, monitoring tools, design software, e-mail, etc.

- Web servers for hosting intranets, public servers, etc.

… and dozens of other enterprise technologies.

Each of the above must be interconnected for the entire organization to function and hence each has the potential to fail in a number of areas. A simple cable disconnection on an international data leased line can cut off every part of the entire organization. Similarly, all equipment can fail in its own unique way.

All IT equipment has the potential to fail or behave erratically, for various reasons. Heavy usage of any such equipment always entails a hidden risk. For example, if the power supply fluctuates there is a high probability of computer disks crashing or corruption of data on many computers. All such IT-related failures, or potential to fail, can be classified as technical risks and sufficient workable, cost-effective alternatives are needed to minimize risk.

What are some of the most common technical risks?

Some of the technical risks common to most organizations are listed below. Disasters in each can range from simple problems to absolute catastrophes.

- Risk to data.

- Virus risks.

- Power failure risks.

- Local area network (LAN) failures.

- Information security risks.

- Telecommunication risks.

- Software risks.

Each of these risks will be explained in detail in a separate chapter:

- Data disasters - chapter 3.

- Virus disasters - chapter 4.

- Communication disasters - chapter 5.

- Software disasters - chapter 6.

- Data centre disasters - chapter 7.

- Information security disasters - chapter 11.

What are some of the most common non-technical risks?

Some of the non-technical risks and disasters that organizations can face are:

- IT staff disasters

- IT vendor disasters

- Reputation disasters

- Financial disasters

- Labour union disasters

- Legal disasters

2: Types of Disaster

- Political disasters
- Natural disasters
- Terrorist disasters

Most of these will be explained in a separate chapter on non-IT disasters – *see chapter 13.*

CHAPTER 3: DATA DISASTERS

'There's no disaster that can't become a blessing, and no blessing that can't become a disaster.'
Richard Bach

This chapter deals with the various ways in which an organization's data can be exposed to risk, and possible prevention methods.

What is data?

A general dictionary defines 'data' as 'factual information, especially information organized for analysis or used to reason or make decisions'. Organizations have various kinds of data in several formats and importance. For example, important finance documents and Excel spreadsheets, computer files in Word documents, databases, E-mails, employee information, customer details, etc, can all be classified as data. Different organizations view data with varying importance. For example, a credit card supplier will consider all details of his credit card numbers, owners, etc, as very, very important. Another organization may view their technical information – software code and the tools they develop – as important data. Irrespective of what it is specifically, data is of paramount importance to any organization and must be protected carefully.

What is meant by risk to data?

A risk to data is a potential for loss or corruption. Every organization depends on various kinds of current and historical data for its business. For example, a simple piece of data can be the details of all customers entered on an Excel spreadsheet and stored on a computer hard disk. A complicated piece of data can be complete details of millions of credit card transactions entered in a huge software database inside a major organization like VISA or MasterCard. Depending on the user, department or organization a loss or corruption of the data can result in various business problems, so protection and safe retrieval of data are of the utmost importance to any organization.

Why and how do companies lose data?

Computer data is a virtual entity. It cannot be protected by security guards or guard dogs. Companies often lose valuable data, for a variety of reasons, eg:

- Many organizations do not invest enough money and resources in taking regular backups or installing proper anti-virus measures. Even today many organizations and businessmen believe that computers can be used and discarded just like any other electrical appliance. They believe that if an old computer fails, they can simply throw it out and buy a later model at a lower price. They fail to understand that regular electrical appliances – refrigerators and fans, for example – do not hold data, but computers do and if they fail suddenly the organization will lose data instantly. Computers should not be treated the same as any other electrical appliance.

- Many organizations do not have proper and qualified technical staff to maintain their computer systems.

- Lack of proper uninterrupted power supplies can also cause disk failures, resulting in loss of data.

3: Data Disasters

- Data is stored anywhere and everywhere – on floppies, CD-Roms, local hard disks, etc. No clearly-identified data storage locations are known to end-users.

- Organizations don't spend enough on anti-virus and other tools – sudden virus attacks can wipe out years of data in a few minutes.

Many business managers think it is easy to hire or replace experienced technical staff immediately from some outsourcing or external company should there be a computer-related disaster. It is not so easy, and is often impossible. It is not possible for an outside IT person, however qualified, to suddenly walk into an organization and start assisting in disaster recovery, business continuity or data recovery. It usually takes several weeks or months for any IT professional to fully understand the IT nature and functioning of any organization. It is not possible to pick it all up in a day, even in small organizations. For example:

A small financial firm had a single computer running a financial application. The application was actually developed by a freelance programmer for a fee. In addition, the freelancer was also maintaining the company data. Only he knew how the stuff worked, how to feed or extract data and reports for the business. Due to a power outage, the central computer disk crashed and the machine stopped booting. Fortunately (or unfortunately) it was a soft crash, meaning that the hardware was okay. Without consulting the freelance programmer, the business owner picked up the phone and called the hardware vendor who had supplied the computer. The vendor promptly sent some new techie to have a look. The techie investigated the issue and realized that the operating system had to be reloaded and informed the business owner. The business owner simply said, 'Go ahead. Do what you want. Get our system back to working condition.' Immediately the techie reformatted the hard disk and reloaded the operating system. The business owner was happy that the computer got fixed without any hardware costs (computer hardware was very,

very expensive those days). Next day, the freelance programmer came to office as usual, and all hell was let loose. The company had lost data worth about two years of data-entry and its huge effort. This is actually a true story. The company had not invested in any backup unit other than a few floppies, which had outdated data anyway. As you must have guessed by now, the techie was none other than the author who had just started his computer career by innocently destroying somebody's data in 1989.

How should organizations store data safely?

Organizations should consider how to store and recover data as soon as they decide to computerize their operations even though it could be used by just one person or a single department. Data backup strategies will have to be created to determine the timeframes, technologies, media and offsite storage of the backups. This will also ensure that recovery point and time objectives can be met. Depending on the organization's size and scale of operations, data storage can be one or more of the following:

- Very small organizations may have just one or two computers. They can have a single directory or folder called 'data' on the hard disk where they create, save and access all data. This data can be as simple as a bunch of Word and Excel documents. As a disaster recovery option, the computer-user can copy all the files every day to an alternative location, eg, floppies, CD-Roms or low capacity USB disks. In the event of a computer failure or disk crash, they will need to call some IT staff to fix the computer, reload the operating system and then copy the data directory back to its original location from the floppies, CD-Roms or USB disk. Nowadays, small pen-sized USB disks with capacities ranging from sixteen megabytes to more than one gigabyte are available for prices ranging from US$50 to about US$400. These USB disks plug in directly to the USB ports on most computers and data can be copied directly to them.

- Small to medium organizations may have several computers. They can dedicate one large computer as a data storage server for all the other users. This is usually called a file server. All users create files – documents or spreadsheets – on their local computers, but save and store the files on specified locations on the file server. For example, the file server can have several directories or folders called 'Finance', 'Design', 'HR', etc. The different departments will have to store their important files into their respective directories. The file server can be attached with a small tape drive that will back up all the specified directories and folders on the server every day. In the event of a file server crash, or an accidental deletion of files, a techie can restore the folders after the server has been fixed. Tape drives are available in various formats, ranging from two gigabytes to more than two hundred gigabytes on a single tape.

- Large organizations will have hundreds, and maybe even thousands. of computers and servers of all types. Servers can range from file servers, database servers, web servers, e-mail servers, and so on. Some sort of enterprise planning is needed to ensure that everyone stores data on properly identified servers. Each of the servers can be backed up using high capacity tape drives, archival systems, mirror servers, etc. For example, huge organizations like Ford, Shell, IBM, etc, will have hundreds of servers of various types containing terabytes of important data. Large investments in backup equipment and qualified staff are required for such massive operations. Depending on the time available, or technical limitations, it can be a centralized backup or a distributed backup.

What are some of the most common storage and backup options?

Various types exist, but there is no single solution for everyone. Some of the common storage and backup options range from inexpensive floppies that can store a couple of megabytes to expensive tape libraries that can hold several terabytes of data. Small tape drives are available for about US$1,000. Gigantic tape libraries that can back up mainframes and large Unix boxes can easily cost more than US$100,000. New technologies such as Advanced ATA, SCSI-RAID, Fiber Channel, etc, are available on modern backup devices. These advanced technologies can back up large amounts of data in short amounts of time. Some of the common storage and backup devices are:

- DATs (Digital Audio Tapes), ranging from 2 to 24 gigabytes.

- DLTs (Digital Linear Tapes), from 20 to 80 gigabytes.

- SDLTs (Super Digital Linear Tapes), 40 to 100 gigabytes.

- LTOs (Linear Tape Open), 100 to 300 gigabytes.

- Tape Libraries of LTO or DLT. These can hold seven to twenty or more tapes inside one box. Combined backup capacities will exceed a terabyte or more.

- Special tape drives for huge machines like large mainframes, AS/400s, Unix boxes, etc.

These devices usually come with their own backup software, or can work with some popular backup software like

Arcserve, Brightstor, Veritas, etc. The manufacturers of these packages also provide various flavours of backup software, eg:

- **Standard:** Basic file backup option

- **Open files option:** This is used to back up files even though some users could be accessing them.

- **Database backup option:** These can be used to back up online databases like SQL, Oracle, etc.

- **E-mail backup:** These can be used to back up e-mail servers like MS-Exchange, Lotus Notes, etc.

- **Remote backup:** These can be used to back up workstations, laptops, etc, which are connected on the network.

- **Image backup:** These can be used to take a snapshot of the entire hard disk, sector by sector. These sorts of backups will be very useful to restore a hard disk perfectly back to its backed-up state in the event of a disk failure.

In addition there are other types of specialized backup options like Norton's Ghost. This software can take a snapshot of the entire hard disk data as a single image file. In the event of a disk crash, the image file can be used to rebuild the hard disk back to its original condition on the same machine. It is also possible to restore the image file on an identical new hard disk on an identical machine model.

What does Internet backup mean?

Nowadays, Internet service providers also provide various options for backing up important data to a server on the Internet, for a fee. Using this service it is possible to copy important data to a secure server or disk space dedicated to an organization. A simple utility or software can be installed on the PC or server that allows the user to schedule backups, select files and folders to be backed-up, password protect files, and so on. Data can also be encrypted for transmission. Though many organizations are currently reluctant to have their data stored on the Internet, it will one day become a popular method of backup once the right security practices are established.

What is a 'geocluster'?

'Geocluster' is short for geographic cluster, which is a very expensive backup option. They are usually used by very big organizations that have to keep data synchronized between different countries. A geocluster is made up of several servers operating in tandem to provide load balancing and fail-over services. For example, if an organization wants to keep a mirror of an important data server data elsewhere in a disaster recovery site in a different city, it can use a geocluster to keep the main and the backup server in sync at all times. Geoclusters are too expensive for small to medium sized firms.

How often should backups be taken, and what should be backed up?

Ideally 'Everything, Everyday'. But, some organizations take a full backup over the weekend and incremental backup daily. As organizations become heavily dependent on computers and data for their day-to-day work, storage and retrieval of data becomes of paramount importance. Companies should ensure that all important data is backed-up regularly and stored in proper fire- and water-proof safes. Organizations should ensure that end-users store their data only on specified file, mail and database servers that get backed-up regularly. In the event of any data loss or accidental deletion, the IT staff can restore the previous day's data back to the user. End-users should also be educated to ensure they do not store any important data on the local hard disks or floppies, etc.

How can one decide what data needs to be backed-up?

This decision should be taken by involving the heads of every department. Ask them what data they consider important and cannot afford losing. Then provide secure folders and other server accesses to the respective department's staff. Educate users to ensure they store their important data only inside the specified server locations. Then back up those locations everyday.

Some organizations back up everything that is stored on a server. This could be a useful practice in some cases. However, it could lead to backups taking a lot of time and unnecessary tape consumption if users store non-business related files like MP3 songs, image files, etc, on the servers.

How and where should backup tapes be stored?

The location is of paramount importance. Backup tapes, like audiocassettes, get damaged easily by heat, moisture, etc. Some of the best practices for storing and using tapes are:

- After every backup, tapes should be labelled and stored in a fire-proof safe in a non-humid area.

- Backup tapes should not be stored inside, or near, the data centre. This is to ensure that the tapes don't get destroyed in the event of any disaster, like fire or water seepage, within the data centre.

- Data tapes are usually stored in an off-site storage that is an alternative site, outside the organization's premises. Also, the same backup tapes should not be used for years and years, as they tend to lose their magnetic retention over time.

- Old tapes should be periodically tested for their ability to restore data. If the tape does not work necessary precautions like taking a new backup on a new tape should be done immediately.

- Old tapes should be destroyed safely so that they do not fall into the wrong hands.

- Implement all the manufacturers' recommendations for the model of backup tape purchased.

How often should backups be tested?

This is a very important exercise. IT departments could be backing-up data over the years but never getting a chance to test whether they can retrieve data from the backup tapes. Tapes don't last forever and get damaged by heat, moisture, disuse, etc, so it is necessary to test every backup tape periodically to see whether you can retrieve the data. For example, if a server fails and if the tape is also not readable then there will be a crisis. It is highly advisable to plan a regular schedule for restoring a sizeable amount of data to a test location from each and every tape that is used for backups. Then end-users can verify whether the data restored to the test location is correct and readable. These exercises can be made part of an organization's DR policy to prevent data recovery surprises.

Will just taking proper data backups daily ensure disaster recovery?

Not enough. Backups on tape or other media will simply ensure that your data is safe. Disaster Recovery is a different ball game. Just having the data on a backup tape will be of no use if the file server blows to pieces. In order to have proper disaster recovery safeguards and recovery methods organizations must invest in the following additional precautions to prevent disasters striking the servers in the first place:

- **Maintenance:** Comprehensive hardware maintenance contracts for all critical servers to ensure the vendor repairs or replaces faulty equipment within hours of failure.

- **Spares:** On-site availability of parts like spare hard disks, spare power supplies, or even a spare machine.

- **Mirror servers:** Depending on low tolerable downtime, some organizations may even invest in having mirrored servers for critical functions.

- **UPS:** Uninterrupted power supply to all critical equipment.

- **Fire prevention mechanisms**

- **Water seepage prevention**

- **Security:** Unauthorized access prevention.

- **Anti-virus:** Virus prevention, with anti-virus updates.

- **Updates:** Applying proper service packs, hot fixes, bios updates, driver updates, etc, as recommended or supplied by the equipment or software manufacturer.

... and other manufacturer's recommendations.

Some questions to ask before starting backups on critical servers and equipment:

- Do we have a complete list of critical equipment that needs to be backed up daily? Have we missed any equipment?

- Do we know what needs to be backed up in each of the above critical equipment?

- Who is assigned to take backups?

- How is the backup taken?

- How long should backups be stored?

- Who is authorized to initiate restores if necessary?

- Will backup tapes need to be stored offsite?

- A few other 'Where', 'How', 'Why', 'When',' What' questions – *see chapter 15* on 'Plenty of Questions' for more.

What do you mean by 'disk mirroring'?

It is possible to have data duplicated in real time across two separate hard disks within a single machine or between two machines. This is called disk mirroring. Disk mirroring ensures continuous availability and accuracy. For example, if there is a server that has two disks of identical capacity, then it is possible to establish a mirror between the two such that data on primary disk-1 always gets mirrored to secondary disk-2. Hence, if the primary disk fails, the secondary disk will have all the data of the primary disk intact. Mirroring can be software-based or hardware-based, although hardware-based mirroring is superior. Nowadays, third party software and hardware is available for mirroring. These packages contain several useful and configurable features not directly available with the basic operating system or hardware.

Can you name some of the high-end storage and backup solutions available today?

Some of the high-end solutions that are available from reputable manufacturers like Veritas and Hewlett Packard are listed below. Visit their websites for detailed information and specifications.

- **VERITAS Cluster Server** is a high-availability solution. It is ideal for reducing both planned and unplanned downtime, facilitating server consolidation, and effectively managing a range of applications in heterogeneous environments. Visit *www.veritas.com* for details.

- **HP StorageWorks Enterprise Backup Solution (EBS)** is a complete enterprise backup / recovery / archive hardware solution built around HP StorageWorks tape-automation products such as the HP StorageWorks ESL9000 Tape Libraries and the HP StorageWorks Ultrium 460 Tape Drive. See *www.hp.com/storage* for further details.

- Replication/data availability solution. In the disaster-recovery arena, VERITAS has a replication/ data availability solution called VERITAS Volume Replicator.

What do you mean by 'database replication'?

Nowadays organizations depend on specialized files called databases that can hold a variety of information in a single computer file. These databases can be accessed and updated by many people simultaneously. Some of the common names in databases are SQL, Oracle, DB2, MS-Access, and so on. Corruption or deletion of a database file can wipe out years of data entered and accessed by thousands of users, so it is vital to take enormous precautions while handling and maintaining databases. Specially-qualified staff called database administrators are necessary to manage databases. A database replication is a partial or full duplication of data from a source database to a destination database. For example, if a database server holds a database called CUSTDATA containing all customer data of an organization, then replication can periodically pump all the data within the CUSTDATA database into another database file called CUSTDATABKUP on a different server. Replication may use any one of a number of methods – synchronous, asynchronous, mirroring, etc. Hence, if the main server fails then it is still possible to extract all data from the backup database server. Nowadays special backup tools are available that can be used to automatically replicate a main database's contents into another database. Various low-end to high-end database synchronization tools are available, with different features available. Some popular examples are available from: *www.dbbalance.com*, and *www.red-gate.com*.

3: Data Disasters

What does 'server load balancing' mean?

Many heavy-duty applications cannot run on just one single machine. There could be thousands of users accessing such an application and the server will get bogged, down unable to service thousands of simultaneous requests. In such cases, server load balancing is necessary. In server load balancing multiple servers are used to host a common application. Through load balancing, traffic can be distributed automatically across multiple servers running a common application so that no one server is overloaded. With this technique, a group of servers appears as a single server to the network. Load balancing can be implemented among servers within a site or among servers on different sites. Using load balancing among different sites can enable the application to continue to operate as long as one or more sites remain operational.

How can one prevent loss of IT equipment?

IT equipment can be broadly classified into two categories:

- Equipment that holds company and user data, like file servers, database servers, hard disks, tapes, laptops, etc.

- Equipment that does *not* hold company or user data, like LAN switches, hubs, routers, monitors, etc.

It is more important to protect equipment that holds data than equipment that does not hold data. However, it does not mean that non-data equipment can be of any less importance. It is just a matter of higher priority, as any company data is of paramount importance to any organization and cannot be purchased from external sources, whereas non-data equipment can be purchased off-the-shelf from several vendors. For example, if an important file server in the finance department blows to pieces, or gets stolen, the situation cannot be resolved just by buying another brand new file server, because data is not re-purchasable. On the other hand, if a LAN switch connecting several machines gets damaged it is possible to buy a new one immediately. Some of the specific precautions for critical equipment that holds data are:

- Have standby power supplies, hard disks or even spare machines if possible.

- Ensure that the equipment is under complete, comprehensive warranty and insurance

- Ensure full daily backups. Insist and verify whether every employee is storing important data only on identified server locations that get backed up everyday.

- Have all manuals, CD-Roms, bootable disks, repair disks, etc, handy

- Verify data integrity regularly by restoring data to a test location.

- Do not store all important data on a single server. Have multiple physical servers to split the load.

- Buy useful recovery tools like Disk Repair, File Undelete, Registry Recover, etc, and become familiar with their usage.

One simple way to ensure that all critical IT systems are covered for various risks can begin as follows.

A critical server housing an important application and data can be protected from predictable disasters by having a checklist:

- System or function name.
- Used for.
- How important is this system for our business?

On-site disaster prevention methods:

- Data from system backed up fully every day.
- Data tapes and storage medium stored properly in fire- and water-proof safes.
- Essential spares like power supplies, spare hard disks, etc, available on-site.
- Servers under comprehensive hardware maintenance guarantee by a qualified vendor backed by an SLA.
- Servers housed in a secure data centre with clean UPS power.
- Servers maintained by qualified and trained staff.
- Servers and data access only to authorized staff
- Servers protected from viruses and hackers by anti-virus and intrusion detection systems.
- Installation of all necessary upgrades, service packs, hot fixes, driver updates, bug fixes, etc, to prevent faults.
- Clear step-by-step documents to assist in data restoration, replacement of spares, etc.
- Insurance to cover theft, fire, damage to equipment, etc.

… and other essential information and precautions.

DR and BC methods

- Hot standby server in a DR or BC site, preferably identical to the one in the main site in all respects.

- Automatic or manual data synchronization between main and standby server.

- Copies of every important document, test plans, etc, placed in the DR or BC site.

- Testing and periodic dry runs at the DR or BC site.

- Other essential information and precautions.

Similar checklists and questions can be asked and tackled for each critical system or business function.

Do's and don'ts for preventing data disasters

Do's

- Ensure that the technical support team is responsible for full and proper backup of all servers daily.

- Invest money in buying good quality tape drives and other backup devices.

- Ensure all important data is stored only on servers that are backed up daily and back up important information daily.

- Learn how to restore data properly.

- Store tapes and key papers in fire- and water-proof safes.

- Test whether you can read old backup tapes and restore data from them.

Don'ts

- Don't allow employees to store business data on their local drives.

- Don't allow unauthorized access to servers and databases.

- Don't allow data to exceed tape drive capacity.

- Don't use the same tapes for a long time.

CHAPTER 4: VIRUS DISASTERS

'Disasters normally don't come alone. They usually bring their family along.'
Anon

What is a computer virus?

A computer virus is a computer program, usually written by intelligent troublemakers (unethical software programmers), to wreak havoc on other computer programs. Viruses come in all flavours. A virus is a software program that serves no useful purpose. It is written with an intention to cause havoc by exploiting some vulnerabilities of the operating system or programs. Some viruses are harmless and can simply pop up with annoying messages, whereas other viruses are deadly and can wipe out all the data on a hard disk. A virus attack can happen in minutes and normally users will not notice the damage until it is too late. Viruses have caused millions of dollars of damage to thousands of organizations worldwide. A computer virus is an executable file designed to replicate itself while avoiding detection. A virus may disguise itself as a legitimate program. Viruses are often rewritten so that they will not be detected. Anti-virus programs must be updated continuously to look for new and modified viruses. Today, viruses are the number one method of computer vandalism.

Example: Virus disaster

CIO: 'Hello Techies? Why are we not able to access e-mail?'

4: Virus Disasters

Techie: 'I think we have a virus attack. Symptoms look like that deadly virus mentioned in the newspapers. It has wiped out all data on all servers.'

CIO: 'But we have anti-virus on all computers.'

Techie: 'We are running an old version. Need the latest version to tackle such viruses.'

CIO: 'We have a major crisis on hand. Call everyone.'

How can you protect your organization from viruses?

In order to protect computers from being attacked by viruses, it is mandatory to have each and every computer protected with an anti-virus software program with regular updates. Several reputable manufacturers like Symantec, McAfee, etc, provide excellent virus prevention and cleaning tools. As of today, there are more than 50.000 types of virus lurking around on the Internet. It is not enough just to install an anti-virus program and hope it will protect you from every type of virus. The anti-virus program must be periodically updated to protect the computer from new types of viruses. Some of the best practices to prevent viruses are:

- Installation of a reputable anti-virus program on all computers (desktops, laptops and servers).

- Updating new virus definitions periodically or as and when the manufacturer provides an update.

- Scanning all machines periodically.

- Preventing users from downloading and installing software programs and files from the Internet.

- Install URL filters: These programs prevent users from accessing unwanted or unauthorized websites. Such tools are available from, eg, Surfcontrol (*www.surfcontrol.com*) and Websense (*www.websense.com*). These websites also offer time-bound evaluation copies to test the application before purchase.

- Preventing users from accessing personal e-mail like Yahoo, Hotmail, Instant Messengers, etc, within the organization as users could receive viruses through attachments sent by unknown persons.

- Scanning every incoming and outgoing mail from the organization. Special tools are available for this, eg, the ones from Trend Micro and Symantec (*www.trendmicro.com* and *www.symantec.com*).

- Educating or preventing users from using floppies, disks, etc, between home and office. This is because a home computer could easily be infected with a virus that could spread through a floppy brought into the office.

- Install proper service packs, hot fixes, program updates, etc, for the operating system and other applications to fix vulnerabilities that could be exploited by viruses.

- Prevent employees from using dial-up connections to the Internet within the organization. Ensure that all Internet access is via proper firewalls.

- Install Internet firewalls with virus detectors.

- Educate users about new viruses and their symptoms.

What is a worm?

Worms are very similar to viruses in that they are computer programs that replicate copies of themselves (usually to other computer systems via network connections) and wreak havoc on a large number of computers within a short time. Unlike regular file viruses that attach themselves to files, worms exist as separate entities. They do not attach themselves to other files or programs. Because of their similarity to viruses, worms are often also referred to as viruses. A well-known example of a worm is the 'ILOVEYOU' worm, which invaded millions of computers through e-mail in 2000.

What is a Trojan?

The Trojan is named after the wooden horse the Greeks used to infiltrate Troy. A Trojan is a program that does something undocumented which the programmer intended, but that the user would not approve of if he or she knew about it. According to some people, a virus is a particular case of a Trojan horse, namely one that is able to spread to other programs (ie, it turns them into Trojans too).

What is a macro-virus?

A macro is a piece of code that can be embedded in a data file. Some word processing software (eg, Microsoft Word) and spreadsheet programs (eg, Microsoft Excel) allow you to attach macros to the documents they create. In this way, documents can control and customize the behaviour of the programs that created them, or even extend the capabilities of the program.

A macro-virus is a virus that exists as a macro attached to a data file. In most respects, macro-viruses are like all other viruses. The main difference is that they are attached to data files (ie, documents like Word or Excel) rather than executable programs. Any application which supports document macros that automatically execute is a ripe target for macro-viruses. One example of a macro-virus is the Melissa virus. It is delivered via e-mail as a Word document attachment with the filename List.doc.

In addition there are several other types of virus. Organizations will have to ensure that they are protected from *all* types.

How can one recover after a virus attack?

In spite of all the precautions, viruses do enter organizations through some loopholes. If the virus does attack, IT support should initiate emergency measures immediately. Some of the common methods are:

- Disconnect and isolate the infected machine from the network immediately.

- Switch off Internet access.

- Run a virus scan and try to remove the virus using various tools and updates.

- Reformat the machine if necessary.

- In extreme cases, it may be necessary to rebuild or restore the last known good backup.

- Switch off other machines on the network to prevent the virus from being spread around.

- Call the anti-virus vendor and try to implement all their technical recommendations to remove the virus.

- Only after you have ensured that the virus has been destroyed should you connect the machines back to the network.

Important Warning: Never ever try to test an anti-virus program by letting loose a live virus in the organization.

Note: It is not always possible to recover from a virus attack. Millions of dollars of data have been lost worldwide because of non-recoverable virus attacks. It may be possible to find a cure eventually for a type of new dangerous virus, but usually the damage would have been done. So, the best method is to take all necessary and continuous precautions and hope that they are sufficient.

How does one update anti-virus on all machines?

It is mandatory to ensure that the latest anti-virus program and its update protect every machine the organization uses. This is actually a Herculean task, depending on an organization's size and the nature of its work. In small organizations, with only a few computers, IT support staff can manually update anti-virus on all machines. However, for large organizations it is not possible to update anti-virus everywhere, every time. It is necessary to use one or more of the following methods:

- Update anti-virus when machines logon to the main server or domain through login scripts, or by installing anti-virus deployment servers.

- Enable automatic update on all machines to connect to a server that holds updates and install them automatically through specialized scripts or batch files.

- Install an anti-virus deployment server that will automatically scan all machines on the network and deploy updates automatically at specified intervals.

- Ensure that all field staff update anti-virus on their laptops by sending periodic reminders and methods to update.

- Implement all the manufacturer's recommendations.

Do's and don'ts regarding viruses

Do's

- Buy a reputable anti-virus program and sufficient licences to cover all computers, laptops and servers.

- Make it mandatory to update every computer with the latest patches and virus fixes.

- Password protect the anti-virus program such that ordinary users cannot disable or uninstall the program.

- Educate users on how to prevent viruses getting into their PCs.

- Scan every diskette, CD-Rom or other data device brought into the organization by outsiders like marketing and sales persons, consultants, etc.

- Virus check every incoming and outgoing e-mail.

Don'ts

- Don't allow employees to bring floppies, CD-Roms, USB keys, etc, from outside

- Don't allow employees to download games, screen savers, utilities, shareware, etc, from the Internet.

- Don't test any live old or new virus on the network just to see whether the anti-virus program can detect and catch it.

CHAPTER 5: COMMUNICATION SYSTEM DISASTERS

'I beg you take courage; the brave soul can mend even disaster.'

Catherine the Great (1729-96)

What are some of the common methods of communications in organizations?

Organizations have come a long way in exchanging information internally and externally from the good old days of plain telephones and telex. Some of the common and extensively used methods of communications are listed below.

- E-mail.

- Internet, WorldWideWeb, Chat, etc.

- Private telephone networks for voice.

- Data transfer using Internet and private leased lines.

- Mobile phones, pagers, SMS, etc.

- Regular telephones, fax, etc.

- Local and wide area Networks.

- Wireless.

… and various other electronic methods.

Organizations have also become heavily dependent on various methods of communication. In fact, many businesses will practically shut down if their communication links fail. For example, a giant online book seller like Amazon.com may lose thousands of dollars if communication links to their website fail even for half a day.

What is a communication failure?

Before the popularity of, and dependence on, the Internet, most organizations worked as isolated silos. The only method of communication to the outside world was through regular telephones, telex or fax. Nowadays, organizations have several methods of communication and are also extremely dependent on them. Some of the common communication methods are:

- Local area network within an office.

- Campus networks.

- Wide area networks between offices and the outside world.

- E-mail (Internal and external).

- Private data leased lines connecting various branch offices of an organization within a city, country or internationally.

- Private voice leased lines connecting various branch offices of an organization within a city, country or internationally.

- Connection via the Internet.

- Virtual private networks via the Internet connecting branches or external suppliers, consultants, telecommuters, etc.

- Wireless communication.

External vendors and service providers usually provide these facilities. Organizations are directly and indirectly dependent on such external sources for their business communications. Communication failures can occur at any time. A disconnect or trouble in any of the above communication modes can cripple an organization's business functions. For example, if an organization depends on e-mail for its support functions to its customers, then it will not be able to provide speedy support to its customers if the e-mail system is down. Each and every method of communication is prone to risks and outages. Hence, organizations must ensure that they have more than one method of communication should the primary method fail for all their critical functions.

Some of the common communication disaster recovery methods or workable alternatives are listed below:

- If the local area network fails, it is possible to have some standby switching equipment to provide essential services to end-users.

- If an organization is connected to its branch offices via a private leased line, it can also implement a slow speed dial up-line in parallel, to be used when the main link fails. Or it can have multiple leased lines that can be used if the main one fails.

- If an organization loses connection to the Internet via a dedicated line, it can have a few dial-up machines hooked to the Internet.

- If an organization's private voice network is down, it can use paid services like IDD (international direct dialling).

- Companies can have telephone services from more than one service provider to avoid failure in one or the other.

- Companies can also have more than one method of standard communication. For example, they can provide customer support through e-mail, phone and fax or by logging on to a website and posting a message.

What are some of the methods for preventing local area network failures?

A local area network or LAN is an important method of connecting computers within an organization. Today, it is not possible to use only stand-alone computers to store or access data. All computers must be linked to access e-mail, send and receive data, access the Internet and so on. Computers are connected using high-end equipment like switches, hubs, etc. High-quality switch boxes are available from several vendors like Cisco, Cabletron, etc. These boxes have ports that can connect just a few computers or hundreds of computers. Some of the common methods of preventing LAN disasters are:

- **Physical redundancy:** Have multiple switches for connecting computers. Switches are available which range from 8-ports to 48-ports or higher. Besides, these switches can be cascaded to provide even more ports. For example, it is better to buy several 24-port or 48-port switches and cascade them rather than buying one integrated switch that has 200 ports. This will ensure that if any switch fails it will affect only a small number of users.

- **Standby equipment:** Have enough standby equipment for critical equipment. For example, have a spare working switch to connect servers and other critical equipment should the main switch fail.

- **Maintenance contract:** Ensure that the switches are under proper and comprehensive maintenance from reputable vendors for speedy replacement during failures.

- **Updating patches:** Install all manufacturer's recommended patches, firmware upgrades, etc.

- **Proper UPS power:** Keep all switches powered by clean UPS power and housed in dust-free, cool enclosures.

What are some methods for preventing WAN disasters?

In today's distributed environment, losing wide area network (WAN) connectivity can cripple a company's ability to function just as effectively as a massive data centre outage. Here are some best practices for designing and developing a resilient WAN:

- **Physical redundancy:** Make sure you have separate cables providing separate services. For example, if you have one long distance service provider providing three WAN links, have all of them on separate physical cables running on different paths.

- **Routing redundancy:** This is a logical redundancy. For example, if the main WAN link between Office 1 and Office 2 fails, configure the network so that the data can be routed via Office 3, if possible. This is called logical switching and is possible with TCIP (transmission control internet protocol) networks.

- **Multiple service providers:** Buy WAN links from more than one service provider if possible.

- **Service provider disaster recovery plan:** Ensure that your WAN service provider has a demonstrable DR plan.

- **Dial-up lines:** Have backup dial-up links connecting all office to be used in the event of a major WAN crisis. Test the dial-up links periodically.

- **Voice connectivity:** Ensure that you have multiple modes of voice connectivity, eg, mobiles, regular phones, IP telephones, Internet telephony, and even old methods of communication – faxes, telex, etc. Do not decommission any mode of communication fully, even if it is old – keep a small working setup for emergency purposes.

- **External consultancy:** Have the WAN tested and certified by some reputable external consultants.

Do's and don'ts regarding communication systems

Do's

- Have more than one mode of communication for every business unit. Implement direct lines, mobiles and e-mail access for all necessary areas.

- Select communications systems and services from different service providers instead of all from one provider. In the event of failure by one, the other channel can be used.

- Ensure your communication links, cables, connections, etc, are secure and tamper-proof.

Don'ts

- Buy all communication services from a single service provider.

CHAPTER 6: SOFTWARE DISASTERS

'The inhabitants of every civilized country are menaced; all desire to be saved from impending disaster; the overwhelming majority refuses to change their habits of thought, feeling and action which are directly responsible for their present plight.' *Aldous Huxley*

What is a software disaster?

For any modern organization to function it will need several computers. But just having computers with only the operating system loaded is not enough. It is not possible to do any meaningful business with just the operating system like Windows 98, Windows 2000, Linux, etc. Special software like MS-Office, databases, e-mail, web software, finance applications, reporting tools, business applications, etc, have to be loaded into computers to do any meaningful work. Third party or external vendors provide most of this software. Nowadays, software is very complicated, and can fail in a number of ways, some minor and some major. For example, organizations normally use various popular databases like an SQL or Oracle server for data. If the whole database is corrupted due to some power problem the organization can be crippled for weeks. Alternatively, viruses can also destroy data in a number of ways. These sorts of problem can be classified as software disasters.

What is a mission-critical application?

A mission-critical application is any software application that is essential to the organization to perform key business functions. For example, any software that handles deposits and withdrawal of funds in a bank can be classified as a mission-critical application. The bank will not be able to operate without this software. Similarly, other applications can also be classified as mission-critical depending on the critical area they are used for. Loss of the mission-critical application may have a negative impact on the business, as well as legal or regulatory impacts. For example, if reputable organizations like VISA or MasterCard lose their critical credit card transactions systems it can be a very big blow in terms of revenue, reputation, legal hassles and other problems.

What are some of the software disasters that can strike an organization?

Software is very complicated, and can fail in a number of ways, so organizations must ensure that every piece of software they use is fully tested under various conditions before they deploy it to production. In addition, a proper change management is necessary to ensure a smooth upgrade (with roll back options) for every new software change. Some of the common types of software disaster are described below.

Operating system-related

Operating systems like Windows, Linux, Unix, etc, are installed and used on millions of computers worldwide. Though the manufacturers take every precaution to provide a stable product, they cannot ensure that it will not fail. Operating systems have various loopholes and vulnerabilities that can be exploited by viruses. Besides, there will be design faults that need to be fixed over time. Hence, organizations should ensure that they use the latest operating systems and the service patches, bug fixes, etc, supplied by the manufacturer from time to time depending on problems discovered. For example, Microsoft has released an operating system called Windows 2000. Based on defects and bugs noticed over a period of time it has also released several hot fixes and service packs that must be applied to eliminate the bugs and defects. All these patches and service packs make the product more and more stable and trouble-free. Organizations that use Windows 2000 on their computers must therefore ensure that they update every machine with the latest patches to protect the machines from design faults. However, the installation of updates must first be tested fully in a test lab and it must be verified that the update will not cause problems with other applications. For example, if a file server running Windows 2000 with Service Pack 3 was having an important finance application, and suddenly the machine was upgraded to Service Pack 4 without testing it might cause the finance application to stop working. This could be classified as a disaster because it might not be possible to revert back to the server's original condition easily. However, if the update was tested in a simulation lab before applying it, the crisis could have been avoided.

Application-related

Similar to operating systems, we have various flavours and versions of other software like e-mail, databases, reporting, development, testing, monitoring, business applications, etc. Each of these packages can have several updates and patches supplied by the respective manufacturer to fix various bugs and design faults that have been noticed. All these patches must also be applied to ensure that the application works smoothly. Proper care must be taken before updating patches and versions on live servers in an organization.

Hardware-related

A particular piece of software or operating system can misbehave if there are hardware design faults in the computer being used. Design faults can be rectified only by the hardware manufacturer and are usually done by providing bios, firmware or driver updates. For example, an ill-mannered SCSI disk driver can corrupt the data being stored on the hard disk(SCSI, pronounced 'scuzzy', means 'small computer system interface' – a piece of software that operates the hard disk). Hence, the disk driver or the SCSI card firmware may need to be upgraded to rectify the error.

The IT support department should periodically check the websites of various products to see if there are any necessary updates and arrange to install them as soon as possible.

What are some of the best practices for software disaster prevention?

In view of the myriad of software programs and tools available, it is essential to manage software implementations properly. Some of the best practices to prevent common software disasters are listed below.

- Ensure that every computer in the organization runs a single type of operating system, office applications and other software tools as far as possible. This will make support and troubleshooting easier. On the flipside, if a software bug, virus or a hacker gets access to the network large scale damage can be done within minutes or hours, so organizations will have to take a balanced approach and ensure that alternative methods are available for each essential function.

- Ensure that every computer runs the same version of hot fixes and service packs. If there are any problems noticed they will be visible on all machines and they can be fixed once, permanently.

- Ensure that every computer is running the latest version of a reputable anti-virus program. Have a periodic and mandatory update policy to tackle new viruses.

- Do not apply service packs, hot fixes, upgrades, etc, without testing them in a test lab.

- Never play around with software settings and features without knowing the consequences.

- Do not allow end-users to install any application or tool they wish.

- Do not allow end-users to download applications, freeware, shareware, tools, DLLs (dynamic linked libraries), etc, from the Internet.

- Do not allow end-users to download attachments from personal mail within the organization.

- Have a proper virus-scanning schedule for every server, desktop or laptop in the organization

- Install manufacturer-recommended bios, updates and driver updates for the computer, eg, video driver updates, disk driver updates, network card driver updates, bios updates for motherboard, etc. These updates usually fix various design or potential bugs and faults discovered by the manufacturer.

6: Software Disasters

- Don't install more than one application or patch at the same time. For example, if a machine has to be updated with Windows Service Pack 4 and also an SQL Service Pack 3, then install each on separate days to see how the machine behaves, unless they need to be installed in a specific sequence to solve some bug.

- Always take a complete backup of necessary data and files before installing any new upgrades or patches.

- Always follow the manufacturer's steps for installation. Do not try to skip any steps or files unless you know its consequences.

- Take the help of professionals and vendors for installations you are not sure about, as there could be several post-installation settings and configurations.

- Do not install any shareware, freeware, hacking tools, fancy screensavers, etc. Some rogue programs can crash systems or cause intermittent problems or data corruption.

- If you are a software development company, ensure that the source code and other applications developed by your programmers are stored or duplicated in more than one location, perhaps many times a day.

- Implement industry best practices and manufacturer recommendations.

- Do not use old and outdated software. Keep upgrading to the latest versions wherever possible.

CHAPTER 7: DATA CENTRE DISASTERS

'WHANGDEPOOTENAWAH, In the Ojibwa tongue, disaster; an unexpected affliction that strikes hard.'

Ambrose Bierce (1842-1914)

What is a data centre?

A data centre is a secure room or rooms where the company's critical servers and other important equipment are housed. A computer data centre is the heart of any modern organization. A disaster here can cripple an entire organization, so special precautions need to be taken to prevent IT disasters, especially within the data centres.

Example: Disaster inside a data centre

Building Security: 'Hello Mr CIO. This is the Building Security Officer calling. Sorry to wake you up at 2 am in the middle of the night. There was a fire in the office just now.'

CIO: 'Heavens. What was the damage?'

Building Security: 'Not much, I think. Luckily the fire engine came within 20 minutes and doused out the fire. The fireman said the fire had damaged only a couple of computers. It didn't spread to other areas.'

CIO: 'That's a relief. Any idea what those computers were?'

Building Security: 'Only those two big black computers in the data centre. The ones with blinking green lights that were labelled

Mainframe 1 and 2. They're burnt to a crisp, along with the cassettes that were stored behind them.'

CIO: 'Mainframe 1 and 2 and our backup tapes? My God! Help!!'

How should a data centre be built?

The simple answer is 'as safely as possible'. Data centres are the heart of any modern organization and should be built with extreme care and conforming to all safety and international standards. A data centre should always be spacious, fire-proof, water-proof, anti-static, ventilated, air-conditioned and in a UPS power-supplied room. Ensure that qualified professionals conforming to all industry standards do the electrical and networking cables inside the room. It should also have proper security and access control facilities so that no unauthorized person can enter the centre.

What are some of the best practices to prevent disasters inside data centres?

Several best practices exist. Today there are organizations that specialize in building state-of-the-art data centres. Here is a long list of common best practices:

- Ensure that the data centre is fire proof. Do not have any material that can catch fire inside or near the data centres.

- Have proper fire alarms, smoke detectors, fire extinguishers, etc, handy. Have fire extinguishers ready and in working condition.

- Ensure uninterrupted power supply (UPS) to all critical equipment. Ensure that the UPS can withstand long durations of main power failures by using generators.

- Never store any important data on machines that don't get backed up. Do not store any data on local drives. Always store them on proper servers that get backed up daily.

- Have an external consultant or fire department inspect and certify the premises periodically.

- Ensure there is no water leakage or humidity anywhere near the data centre.

- Ensure that the data centre is not directly above or below any hazardous areas. For example, the data centre should not be located above a kitchen, or a place where they store fuel.

- Do not allow unauthorized persons to operate any equipment or enter the data centre.

- Have a proper air conditioner to cool the IT equipment.

- Do not have any plants and other decorative material inside the data centre.

- Do not store empty cartons, packing material, inflammable material, liquids, etc, inside the data centre.

- Have static eliminators installed at key locations. This will prevent electrostatic charges from gathering around carpets, doorknobs, etc. Static electricity is very dangerous and can cause minor to major shocks to anybody touching a charged object. In addition, they also cause electronic equipment to fail.

- Do not remove the chassis cover on any equipment without powering it down. Attach a static discharge belt for additional safety. Some high-end servers allow hot pluggable hard disks that allow you to install a hard disk when the system is running. Unless it is absolutely necessary to keep the system powered on always, do not try to use such features. Shut them down and then upgrade or repair.

- Do not draw more power than recommended from any power supply outlet. Have proper fuses and circuit breakers installed.

- Ensure the data centre is free from pests, rodents, ants, etc.

- When lifting or relocating equipment ensure that it is done by qualified personnel using proper tools like trolleys, rubber mats, etc.

- Do not stack equipment one above the other. Ensure adequate ventilation and space for easy maintenance.

- Don't touch any IT equipment with wet hands. Do not touch the motherboard or other electronic components when the chassis is opened for any reason. Static electricity on the hands and fingers can destroy electronic chips.

- Always use high quality electrical and computer cables, even if they are expensive compared to ordinary cables.

- Replace servers, computers and software with the latest and better systems periodically. Most manufacturers declare certain models and versions obsolete within two or three years and stop supporting them. If your organization is still using old systems or older software you run the risk of not getting any vendor support when they fail.

- Keep all manuals, original disks, CD-Roms, licence numbers, etc, in a safe documented fire-proof place. Keep multiple copies of every important document. Keep an additional copy in the disaster recovery centre.

- Ensure that the server passwords are used only by authorized personnel and change them often. A critical server's password with an unauthorized or incompetent person is a disaster waiting to happen.

- Delete all unwanted user IDs periodically to prevent unauthorized use.

Other precautions to prevent IT disasters

- Always buy branded and reputable hardware models even though they may be slightly expensive. Branded and reputable manufacturers invest in the necessary R&D to ensure that their products are tested and supported fully. Unbranded and assembled computers are normally a mixture of several unknown vendor products, and usually have freak problems that will get no support from the equipment seller.

- Ensure that your technical staff undergo periodic training in the latest areas of IT support.

- Periodically reduce the number of unwanted and old IT machines throughout the organization.

- Always buy tested and proven versions of software and hardware. For example, if a manufacturer releases a new version of an operating system or office application don't try to install it everywhere immediately just to be ahead of the rest. Wait for the software to stabilize in the market and then implement it in a staged manner. Remember that most software and operating system upgrades are one-way processes and cannot be rolled back easily. For example, if you suddenly decide to upgrade to the latest operating system and later discover that none of your critical finance applications work properly this could lead to a serious crisis.

CHAPTER 8: IT STAFF DISASTERS

'There are men in the world who derive exaltation from the proximity of disaster and ruin, as others from success.'.

Winston Churchill (1874-1965)

Who do you mean by IT staff?

Every modern organization will usually have several staff or departments (internal or outsourced) for maintaining and troubleshooting the IT infrastructure. Such staff or departments are usually called IT staff, tech support, and technical assistance, etc. They usually have specialized training and the skills necessary for maintaining critical IT equipment. For example, there could be a specialized team just to manage backups and restorations of various servers in the organization. They could be trained in using the backup software, how to back up, what to back up, how to restore, etc. Or there could be a dedicated team just to manage and operate the company e-mail systems.

What are the general precautions to prevent disasters relating to IT staff?

The IT staff of any organization can usually be categorized as key staff, because they handle the organization's critical IT equipment. A people-related disaster like resignation, injury or death to one or more critical IT staff could paralyze an organization. Ironically, it is possible to replace an organization's CEO overnight, but it is nearly impossible or highly risky to replace or lose key IT staff suddenly.

Some of the common precautions to prevent IT staff-related disasters are listed below.

- Don't have all your IT staff seated in one place. For example, don't have all your IT staff working in the same location or building. If something happened to that building, all the knowledgeable staff would be affected, and it would cripple your ability to get the network up and running again.

- Ensure every IT staff member is adequately trained in all or most support services. For example, it would be risky to have just one person who knows how to operate the backup software or who has the admin passwords for all servers. If that person quits or meets with an accident then nobody else can take system backups or perform administration activities.

- Pay industry standard or better salaries to IT staff. Good and competitive salaries ensure low resignations and attrition. Many organizations still believe IT staff are a dime a dozen. Many business managers think that they can always hire or replace experienced technical staff immediately from some outsourcing or external companies should there be a computer-related disaster. Believe me, it is impossible. It is not possible for any new IT staff, however qualified, to suddenly walk into an organization and start assisting in disaster recovery, business continuity or even plain day-to-day support immediately. It usually takes several weeks or months for any new IT staff to fully understand the IT nature, functioning and the culture of any organization, its past history, operation of legacy systems, etc. Even in small organizations this isn't easy.

- Have an adequate staff ratio: *see below*.

- Don't hire temporary staff just to reduce costs. Temporary staff will usually have no commitment or loyalty to the organization, and will always be on the lookout for better opportunities elsewhere. Secondly, they may leave at any time with very short notice, causing cause serious IT service issues to any organization.

- If you are outsourcing IT staff ensure that you demand a minimum set of IT qualifications and experience from the staff being supplied by the outsourcing vendor.

What is an IT staff ratio?

In order to maintain a large IT infrastructure, it is necessary to have a sufficient number of IT staff to properly manage various systems. Irrespective of the amount of automation, automatic systems, etc, enough qualified staff are still needed to understand, control, manage and run the operations. However, many organizations fail to understand this important aspect and try to keep the ultimate minimum number of staff, or a slave-sized team, to maintain a large IT infrastructure. The standard decision factors are cost saving, businesses unable or unwilling to invest in more headcount, etc. However, managing a large IT infrastructure will put enormous pressure, stress and overheads on the staff if the IT department is too small. Most IT staffs are struggling to meet service expectations that are too high for the current sizes of IT departments. Naturally, this will result in frequent resignations, improper process compliance, delay in support, and other issues that will slowly engulf the organization. The revenue loss due to an overloaded, understaffed IT service team will be several times the saving in salaries of having twice the number of IT staff.

It is not enough for organizations to say they have implemented IT best practices just by preparing a bunch of process documents, procedures, policies, etc – they also need the right number of staff to practice IT in its recommended way. Secondly, there is no point in committing high levels of availability everywhere when there is a shortage of staff to maintain even basic services. This is where staff ratio will help.

An IT staff ratio means having the right number of IT staff for a certain number of end-users and IT equipment. For example, a general rule of thumb is to have two IT staff to support 100 end-users using 100 computers and about three or four servers. However, it would be unreasonable to have the same two IT staff continue to support the organization when the strength grows to 200 end-users. Naturally, the IT staff will have to increase in direct proportion to the end-user count. Many business managers may argue that it is possible to simply implement a few fancy tools and not increase IT staff. However, such arguments usually do not work out in real-world, practical scenarios. Fancy tools are usually very expensive, and will anyway need highly-qualified staff to operate and maintain, plus there will be ongoing costs. Hence it is absolutely necessary for businesses to ensure that they have the correct IT staff strength to maintain the expected levels of availability.

Example: Inadequate staff ratio affecting business

As you may have observed, computers always mysteriously fail at the most inappropriate time. For example, a computer could fail during an important business presentation to a potential client. If there were enough IT staff, a techie could speedily attend the trouble within minutes to give some immediate work-around. This can give a good impression to the client about the organization's service standards. On the other hand, if there were a shortage of staff and the techie walks in after two hours (or does not turn up at all) it could lead to an acute embarrassment plus an abrupt end to the meeting, or losing the client (a very high probability), tempers rising and so on.

There is no magic number for the IT staff ratio. Organizations will have to gather the following statistics and arrive at an optimum number:

- Average number of end-user calls per day

- Average number of call backlogs per day

- Call response time compared to committed time

- User downtime

- User downtime calculated in financial terms

- Growth of end-user count

Companies that wish to compete based on properly fulfilling commitments made to their external and internal customers must invest in correct IT staff to end-user ratios to remain competitive. Otherwise, the company could slowly suffer from an internal decay that could cripple the business.

What are the usual reasons for IT disasters?

Many organizations have implemented computers, software, telecommunications, etc, for running their businesses. However, these implementations are usually done without proper planning of any sort. This example shows how most organizations usually implement IT in their organizations:

Example

A small industry's business owner may buy a single computer initially for general use. After seeing the benefits of using computers, he may immediately decide to buy 25 more for his staff. Within a short time his business will be computerized, and very soon IT support headaches will enter the business. Using a computer may be easy, but maintaining a computer system is a complicated task. Users may suddenly experience crippling virus attacks, equipment failures, software licensing issues, data corruption, data loss, upgrade issues, and so on. They may not be in a position to support and maintain a computer network and its associated functions. Overnight, a smart purchasing assistant may undergo a crash course in computer maintenance, or buy a book called 'Computer Maintenance for Dummies', and soon will be given the responsibility for technical support of the business along with their other responsibilities. This is how IT departments start in thousands of organizations. However, this sort of approach will soon lead to major and uncontrollable issues later.

Some of the common hassles faced by many small and even large organizations are listed below:

- Roles and responsibilities of staff are not clearly defined or non-existent.

- A single IT staff or a very small team of IT staff responsible for anything and everything related to IT.

- Lack of clearly defined and simple processes. No service level agreements, vendor agreements, technical training, etc.

- Business and technical staff not seeing eye to eye. Poor management buy-in, inadequate funding, culture issues, resistance to change, etc.

- Businesses not understanding essential factors of using IT in their organizations like having proper IT staffing, exponential hardware and software budgets, on-going costs, frequent and mandatory upgrades, etc.

- Technical staff concentrating only on technical matters, and unable or unwilling to understand business needs.

- No structured customer support mechanism. No help desk or service desk facilities.

- No proactive IT trouble prevention methods. Only reactive support. Troubles get solved after it occurs with no prevention mechanism in place.

- IT staff using outdated tools and equipment due to various reasons resulting in the IT department out of sync with modern business demands.

… and several more.

What are some of the best practices to be followed by IT staff?

Proper IT service is a very important aspect of any IT department. Many organizations do not have any good processes in place to manage IT services. Different organizations follow their own proprietary methods to provide internal IT support, but there are industry standard practices readily available that can be easily adopted by any organization of any size. One of the best-known is the ITIL Practices, also known as the *IT Infrastructure Library*. ITIL was prepared by the OGC (Office of Government Commerce, UK) and defines best practices in IT service. Excellent books, written by the ITIL gurus, are available on the subject. Visit *www.itil.co.uk* or *www.itsmf.com* for further information and details of books on the subject.

What are the main benefits of using ITIL?

Many organizations believe they have already implemented excellent self-developed IT services and don't need any change. That might be right. But on closer examination it is more likely to be found that they are missing out on various processes that could enhance their IT department. The benefits of using ITIL are simply enormous, eg:

- Proven and tested processes. No need for businesses to re-invent the wheel for implementing IT services in their organizations. Covers end-to-end.

- Improved quality of IT service for business functions.

- Reduced downtime, reduced costs, improved customer and end-user satisfaction

- Measurable, controllable, recoverable.

- Proactive rather than reactive. Clearly-defined roles, responsibilities and activities.

- Greater understanding of IT and its limitations by the business. Business will understand IT better.

- Continuous improvement, stability, and problem prevention.

- Improved business image. Businesses will also learn what to commit, and what not to commit, to their external customers.

8: IT Staff Disasters

How can change management prevent disasters?

Most modern organizations have implemented change management procedures for technical implementations within their organizations. What this means is any changes like additions, deletions, modifications, replacements, etc, to any part of the IT infrastructure must go through a series of approvals and sign-offs before the changes are actually implemented. For example, the management should not allow any unauthorized technical changes to the infrastructure. A knowledgeable change management team will need to study the change requested, and view it from several technical and non-technical angles before giving the go-ahead. Having a proper change management process can prevent several types of disasters, for example:

- Preventing any IT or network changes during critical periods. For example, organizations that sell consumer goods over the Internet or through retail stores should not disturb their IT infrastructure (on which they depend for sales) in any way during the Christmas period. Suppose the IT staff install an untested software patch just before Christmas on an organization's online sales web server. If the patch misbehaves and the server crashes due to a bug, customers cannot purchase the company's products during such a critical time, causing loss of reputation and other negative impacts.

- Businesses should not allow any maintenance activities (except emergency fixes) on production systems during business hours. Not allowing any technical changes to be done during business hours can prevent any unexpected disasters and business disruptions.

- All IT changes must and should have a proper back out plan. For example, if you are upgrading some software on an important server, then an accurate snapshot or baseline must be taken before installing the upgrade. If the upgrade fails or causes some other unexpected problem, the system can be reverted back to the previous baseline. Tools like Norton Ghost or Disk Imaging software can help create accurate images or snapshots of the systems being upgraded or modified.

What are the other risks relating to IT staff?

Risks and disasters can happen with every employee. However, risks from IT staff can be more severe as they are specialized employees who may have complete access to all critical equipment. They will have access to equipment and data that even the CEO will not have. Of course, organizations cannot survive without having some IT staff, but care can be taken to minimize the risks relating to them. Some of the common risks are listed below.

- A disgruntled IT person can be an enormous threat to an organization. He or she can simply destroy data from critical equipment for revenge.

- Employee dissatisfaction among IT staff resulting from lack of growth opportunities, inadequate salaries, overworked/underpaid situations, etc, are all potential threats to an organization.

- An inadequate IT staff ratio is also a potential risk and a disaster waiting to happen. Not having enough staff can gradually reduce an organization's capability to be competitive. Problems will get fixed slowly, processes will not be followed, dangerous IT shortcuts will become commonplace, data backups may not be regular, etc. All these can lead to disaster sooner or later.

- IT service is a serious business and should be handled by mature and responsible staff with at least several years of proven experience.

- Organizations should also ensure that the IT staff have no drink problems, if possible:

Example

A highly qualified techie was managing a certain large organization's IT infrastructure. However, the techie had an alcohol problem, and would usually get completely drunk every evening. During a critical project implementation there were some technical problems late in the night, so the project chaps called the techie to come over and solve the problem. Unfortunately, the techie was very drunk by then but he somehow managed to crawl into the office. Unable to understand what was going on he picked a fight with some project members and started banging on the keyboards and terminals. Luckily, no serious damage was done to any data or equipment. Finally, building security had to be called to handle the techie.

As you can see from this example, IT staff who have drinking problems can be a great threat to any organization.

- Resignation by critical IT staff: In this competitive world, qualified and experienced staff are always in high demand everywhere. Hence organizations should ensure that they retain their qualified IT staff as far as possible, and also have enough qualified IT staff to handle sudden resignations or employee 'poaching' by the competitors.

CHAPTER 9: IT VENDOR DISASTERS

'What you spend years building, someone or something could destroy overnight. Build anyway.'   *Mother Theresa*

Who is an IT vendor?

All organizations depend on a number of external and third-party agencies for hardware, software, telecom, support, consumables, spares, and other IT equipment. It is not possible to run any organization without having one or more IT vendors supporting some critical equipment or function. Selecting the right vendor is therefore of utmost importance. For example, if an organization is heavily dependent on e-mail for its business, the vendor who supplies and supports the e-mail software will be very critical to the organization's business. If the e-mail vendor goes out of business then the organizations that have implemented those e-mail systems will not get any more upgrades or support.

What is an IT vendor-related disaster?

A disaster occurring to a critical IT vendor is indirectly a disaster for any organization using the vendor's products. For example, assume an organization has purchased a database application from a vendor, XYZ Databases Corp, to load all critical financial and other information. If XYZ Databases Corp goes bankrupt, gets hit by some internal disaster or goes out of business all organizations depending on XYZ Databases will be affected. Suddenly, there could be nobody to support, upgrade or troubleshoot the application. These can be classified as IT vendor-related disasters.

How can organizations protect themselves against IT vendor-related disasters?

Organizations usually have no control over disasters relating to IT vendors, but they can minimize the effect by having more than one vendor for similar functions wherever possible. For example, organizations can buy hardware and networking equipment from different manufacturers and vendors or they can have equivalent software from multiple manufacturers for similar functions. In the event that an IT vendor does go out of business, organizations should be in a position to speedily switch over to equivalent alternative systems from other sources. Some of the key factors to be taken into account when choosing vendors are:

- Reputation of the vendor and manufacturer

- Availability of competitive products

- Disaster recovery competency of the vendor

- Availability of support from third party sources

How does one prevent IT-vendor support disasters?

Most IT vendors just supply hardware and software. Some of them also provide basic support like first-time installation, troubleshooting, etc. A few of them provide detailed support and consultancy options. However, organizations should not depend entirely on one IT vendor for all advice and support. They should also have qualified and knowledgeable internal staff to verify and understand the business pros and cons of vendors' recommendations. A vendor may not appreciate the business implications of his recommendations. Organizations must understand that not all vendors may be qualified to give accurate business protection advice. For example, a vendor may view a hard disk crash and its replacement on a critical server as a simple issue, whereas it may be seen as a bankruptcy signal for the business owner if that disk was not being backed up regularly.

Example

In another potential disaster case, there was a certain Novell Netware file server that had developed some freak freezing problems, and would often lock up. The server was the heart of the organization, and important files were stored on it. The IT staff within the organization did not know how to solve the problem without losing data, so a vendor was called in to have a look. To everyone's shock, the vendor actually suggested reformatting and re-loading the Novell operating system to see if that would solve the problem. He did not understand the business implications of suggesting reformatting or reloading on a live server containing valuable data without a bunch of prior precautions. Fortunately, reloading was not done and applying some minor upgrade files from the Novell website later solved the problem.

Should IT staff be outsourced?

Yes and no. It depends on the business' management and how they view IT. Nowadays thousands of small to large organizations outsource their IT functions. Many view IT as a burden they can avoid and try to outsource it. Sometimes outsourcing is actually done with a herd mentality attitude – everybody is doing it, so we should too. Other organizations view IT as a core, essential function that cannot be outsourced for a variety of reasons. Either way, there are risks. Actually, for best results, IT should be a combination of internal employed staff and some outsourced staff. This is because many organizations become over-zealous and decide to outsource every IT function. This is when trouble starts. There will be nobody within the organization with the required technical expertise to verify or certify whether the outsourcing company is actually delivering what they have committed or promised. Although there are some cost advantages with outsourcing, a balanced approach has to be taken after considering various factors, eg:

- Outsourcing decisions are usually based only on cost factors. Hence, the cheapest quote from a bunch of vendors will get the order. Six months down the line, or during renewal, if another vendor quotes US$50 less he will get the order.

- Vendors may or may not have the expected loyalty, dedication and commitment to a serviced organization's business functions.

- Information security could become a serious issue.

- Inadequate service level agreements (SLAs) – or none – can cause painful legal problems when there are more important IT service issues to be worrying about.

- Outsourcing vendors usually rotate staff between different companies and smooth transitions are rare. For example, there could be a disconnection between the outsourced IT staff that were providing support between January to March, versus the new outsourced staff who will provide support between April to June.

Example

A certain Company A had outsourced all its computer and network services from a reputable networking Company B. The outsourcing contract was for one year. About half a dozen of Company B's tech staff were fully involved in implementation of the computer and network equipment for Company A, so only those IT staff knew everything related to the IT infrastructure. After six months, Company B wanted to recall its staff to use them for another client and offered to provide a different set of staff to Company A. The new staff had to start everything from scratch. They were not familiar with the setup, staff names, priorities, etc, and started learning afresh on the job, resulting in total chaos to Company A. Later, Company A finally ended up hiring the previous IT staff of Company B at much higher salaries than they were being paid by Company B.

- The best staff of the outsourcing company will usually be placed in the best-paying client's premises.

- Information security, confidentiality, etc, will become a serious issue as non-company staff will have access to internal information.

- During a change-over of outsourcing vendors, the hand-over of responsibilities from one vendor to another will always be a serious and troublesome issue. The outgoing vendor will usually not do a proper handover to the next vendor, as the account does not matter to them anymore.

- Outsourced IT staff usually go strictly by the book or scope of contract, and will rarely be flexible without additional costs. For example, if there is a need for the outsourced IT staff to be present in the company after office hours or weekends for some urgent work, it will usually involve additional hourly costs.

- Other factors, like location of the outsourcing company, travel distances, holidays, internal problems, etc, all affect a client organization.

9: IT Vendor Disasters

What can be outsourced?

It depends on the nature of the organization and the availability or non-availability of certain skills in-house. Other factors like costs, logistics, security, etc, also play an important role. Usually defence and military establishments do not like to outsource anything and expect to have qualified in-house personnel to handle everything, except for certain functions that have no security impact. Some organizations may like to outsource everything as they feel they don't have, or can't afford to have, expensive, qualified IT staff on their payrolls. The following areas of work are outsourced in many companies today:

- **Desktop and server hardware support:** An organization's internal tech staff may not have the necessary skills to repair or replace various types of failed or new hardware. This can be outsourced and will mainly involve repairing or replacing failed hardware, setting up new hardware, etc. Depending on the speed necessary, spares and IT staff can be external or housed on-site.

- **Networking:** When organizations grow or set up new offices, factories, etc, the entire place has to be wired with data and voice cables necessary for local and wide area networking. It does not make sense to have qualified in-house staff with those skills and the job can easily be outsourced to reputable vendors to wire up a building.

- **Turnkey projects:** Assume that an organization wants to establish a branch office in a different location or city. The entire project of cabling, networking, installation of new equipment, power, etc, can be outsourced to a reputable organization who can complete and hand over the project for a fixed fee within a fixed timeframe.

9: IT Vendor Disasters

Whatever the reason for outsourcing, or not, organizations must consider the availability of critical support, spares, talent, etc, required for ensuring disaster recovery and business continuity. Clear service level agreements outlining a detailed scope of work, expectations, roles, responsibilities, etc, must be enforced to cover all preventable risks.

Some questions to ask vendors

Here are some questions to ask, and get satisfactory answers to, when selecting vendors for critical equipment or services. When one or more vendors go bust, you must be able to quickly locate another vendor to maintain the service to your end-users as quickly as possible. It is always better to have more than one vendor for any product or service.

- Does the vendor have enough trained support personnel to handle technical support?

- Does every support person carry a mobile or a pager for contact during emergencies or otherwise?

- Does the vendor have adequate stock of critical spares?

- Does the vendor have a 24x7 support option?

- Does the vendor have a DR or BC plan?

- Can the vendor give some good references to verify and/or any other testimonials or certification?

Is it necessary to have contracts with vendors?

Absolutely. If you are using external vendors to support or maintain critical equipment and services it is absolutely necessary to have a proper contract signed and agreed by both parties. The contract should be prepared in detail, covering the following:

- Scope of work

- Exclusions

- Roles and responsibilities

- Service hours

- Duration of contract

- Spares support

- Reports to be provided

- Payment terms

- Penalties for non-adherence

Contracts should be prepared with the support of staff from technical, financial and legal departments so that all aspects are properly covered and worded accurately. A contract must withstand scrutiny by lawyers or the courts, if necessary. In addition, a detailed technical service level agreement (SLA) is also necessary to ensure proper support. Periodic audits should be conducted to see that SLAs are being met.

What are the key elements of a maintenance contract or an SLA?

As mentioned before, it is necessary to have proper, written agreements with appropriate vendors, service providers, consultants, etc, responsible for maintaining critical services for an organization. Without a clear, signed agreement it is not possible to ensure or expect that the required assistance will be provided by external parties for essential activities in various situations.

A general purpose SLA will normally cover the points listed below. Each point needs to be elaborated in clear and definitive terms for the area of coverage. Additional items can be added depending on the specific nature of work or industry.

- Name of the project or area of support.
- Contract number or reference number with date.
- Start date and end date for contract.
- Description of the project or work expected.
- Parties to the agreement, including authorized persons, departments and workplace addresses.
- Detailed scope of work.
- Common obligations of both parties.
- Out of scope (both parties).
- Assumptions, constraints, risks and limitations.
- Hardware, software, spares, other requirements.
- Legal aspects, jurisdiction, non-disclosure clauses.
- Financials, budgets, payment terms, penalties, additional costs, extra charges, taxes, billing methods, etc.
- Standard working hours or service windows covering number of hours per day, holidays, etc.
- Number of staff required on-site or on call.
- Training requirements.
- After-hours work, eg, weekend work, if any.

- Help desk or support procedures, turn-around times for response, resolutions, work-arounds, etc.

- Incident and problem management procedures.

- Escalation procedures.

- Change management procedures.

- Reports and metrics (what standard reports will be exchanged).

- Project termination clauses, notice periods for closure.

- Signatures of authorized representatives from both parties.

Example: An IT service *without* a maintenance contract

IT Support: 'Hello ABC Computer Company? We are calling from RockSolid Corp. One of our main server's power supplies has failed. Can you replace it immediately?'

ABC Company: 'Can you tell me the serial number of the server?'

IT Support: 'It is QW1246.'

ABC Company: 'Sorry, that server is out of warranty, and also not under any support maintenance agreement, so we will not be able to replace the power supply.'

Example: An IT service *with* a maintenance contract

IT Support: 'Hello ABC Computer Company? We are calling from RockSolid Corp. One of our main server's power supplies has failed. Can you replace it immediately?'

ABC Company: 'Can you tell me the serial number of the server?'

IT Support: 'It is QW1246.'

ABC Company: 'Thank you for the details. That server is under our maintenance contract. We will replace the power supply within the next four hours.'

CHAPTER 10: IT PROJECT FAILURES

'Mr Corleone is a man who insists on hearing all bad news immediately.'
The Godfather

What is an IT project?

Modern organizations today require myriads of IT equipment like computers, telecom devices, data and voice lines, security devices, firewalls, software, etc. Proper selection, installation, configuration and maintenance of those IT environments are of crucial importance. The proper implementation of such equipment can be considered an IT project. Dozens of factors must be considered during an IT implementation. Some of the common, and most important, factors to be considered in an IT project are:

- Proper selection of the right equipment

- Design and capacity planning

- Cost factors (one-off and ongoing)

- Inter-dependencies

- Training and support requirements

- External and other factors

- Miscellaneous issues

Many companies outsource their internal IT work, and may also serve as service providers to external clients. Organizations are heavily dependent on external service providers to implement, maintain and support their IT infrastructure. They also outsource entire IT projects to external service providers and consultants. For example, a number of industries depend on external hardware vendors and software firms for developing and implementing IT projects. IT projects can fail in a number of ways, however, and such failures can affect an organization severely, eg:

Costs: Improperly designed, or poorly-chosen, equipment can result in a lot of wasted costs. Sometimes more money is wasted by not killing an improper project on time.

Reputation losses: Failure to implement an IT project that meets a customer's requirements can cause reputation losses for the service provider as well as losses for the client. For example, poor project deliveries to customers can affect the reputation of an organization, causing loss of new or existing customers.

Legal issues: Clients may also sue service providers if they goof up critical projects. For example, external customers may sue software development companies for delaying or messing up the implementation of a critical software project.

These are all disasters from a business perspective.

Why do IT projects fail?

As with any project, there are a number of risks associated with an IT project. These days IT projects are more likely to affect a business than non-IT projects. This is because organizations are more and more dependent on IT for all their data, communications, connectivity, etc. Secondly, organizations have to constantly upgrade and implement newer and newer technologies to remain competitive.

Implementation of such technologies will normally involve creating and managing IT projects, but there are many reasons why such projects fail, eg:

- **Poor capacity planning:** For example, some new multi-user data entry software may not be able to handle heavy data entry loads. Or, a new network connectivity between offices may not be able to handle the level of data traffic between the sites.

- **Bad technical designs:** It is possible for an entire IT project to have a bad technical design. For example, service providers or IT departments may select wrong or unsuitable IT equipments due to budget constraints, political reasons, insufficient knowledge, sales gimmicks, fancy marketing brochures, etc. The entire design may be technically flawed or may not be the right solution for the need. In such cases, the technical implementation may be successful but may not meet business requirements later. In other words, it leads to an 'Operation successful, Patient died' situation.

- **Political reasons:** No organization is free from internal politics and back-stabbing of various sorts and degrees. In addition, there are more culture clashes these days as workforces become more and more culturally diverse. A project initiated by the head office located in one country may cause job losses to employees situated in another country. Such projects have the risk of failure or delay caused by employees of service providers who may be negatively affected.

- **Budgets and timelines:** Projects can also fail to meet business expectations due to inadequate budgets or unrealistic timelines. It is a well-known fact that most projects are given insufficient budgets and timelines by the management. Secondly, many projects have inaccurate estimates to begin with. This is because of the way most IT projects are sold to senior management within organizations. Tell the truth in many organizations and projects don't get funded, so project managers and sponsors lie instead to buy time and somehow start the project but cause dogfights and bloodbaths later.

- **Bad managers:** According to various workplace surveys, studies, etc, bad or weak managers account for a large percentage of project failures. The badness can be in terms of abusive behaviour, poor knowledge, inadequate experience, lack of professional guts, etc, leading to various kinds of workplace and team disorders. Bad managers can cause projects to fail by causing resignations of key talented staff, unachievable deadlines, over-commitments, etc. Believe it or not, many employees prefer not to report potential problems to senior management, for obvious reasons or because of past bad experiences. Even today many managers do not like to hear bad or expensive news. Very often a 'shoot the messenger' attitude is displayed, instead of appreciating the gesture. So employees avoid or delay telling bad news to them, resulting in catastrophes later.

Real-life example

In a particular organization, electrical fuses used to blow often due to erratic voltage and other uncontrollable issues. So the electricians of the firm frequently used ordinary wires instead of proper fuses. One day the entire system was destroyed. The electricians knew they had to use proper fuses, but did not use them. The reason was that those electricians were reluctant to deal with a foul-mouthed supervisor in charge of buying fuses and other essential parts.

- **Business decisions:** In many medium- to high-profile projects the objectives, scope, budgets, timelines, people, etc, are decided at very senior management levels. Next such projects will be forced downwards with a tremendous amount of pressure, clout, secrecy, veiled threats, etc. Very often, the top guys will refuse to accept or listen to any real world issues or specific problems that might affect the project. Frequently the messengers of bad news are punished or pushed out of the projects leading to a fear psychosis, covering up of bad news, cost escalations, etc. Often bad news identified at the lower levels gets sugar-coated and presented as non-issues, or even good news, to the top management. Soon reality kicks in and it will be too late to prevent the impending project failures.

- **Organizational inadequacies:** Not all organizations are capable of flawless project execution within agreed costs and timeframes. It doesn't matter whether the organization is big or small. Sometimes big organizations may not be able to fulfil a small organization's needs because of the inherent way in which big organizations function. Long timeframes, elaborate processes, lavish overheads, etc, may cause lot of grief to small organizations. Conversely, a small organization may not have the resources and bandwidth to handle IT projects for a large conglomerate.

- **External factors:** Projects can also fail due to external factors beyond the control of the company, eg, political disturbances, vendor issues, regulatory issues, etc.

How can organizations avoid IT project failures?

In spite of significant progress on project management methodologies, best practices and the availability of new technologies, the success rates of medium- to large-scale IT projects are still poor. It has been observed that fewer than 50% of all such projects are actually successful in a true sense. The rest are either cancelled completely, delayed beyond acceptable limits, over-budget, or go completely haywire and do not meet even the minimum business or functional requirements.

IT projects, as the name suggests, are always technology-related, and can be simple or extremely complex. The quirk of IT projects is that if the projects are managed by pure technical experts they could turn out to be a great technical success but may not meet business expectations. On the other hand, if a pure business manager runs the project, it can lead to a technical failure caused by lack of expertise. It is often difficult to get a balance between the two. This is where the role of a CTO or a CIO is most important. He or she must interface between the business and technical experts for IT projects.

Some of the ways to avoid IT project disasters are:

- **Customer requirements:** The heart of any IT project lies in clearly understanding and documenting the customer's requirements. The requirements must be agreed in writing and clearly signed-off. More often than not, customers may not be very clear as to what they want. They may just give vague or superficial requirements, but expect something more dramatic after implementation. The trick is to get the correct requirements from the customer before the project is started. New requirements during the course of a project must also be captured and properly processed.

- **Commitments:** In the face of reckless competition many organizations over-commit on their product or services.

Such commitments will normally not be achievable due to real world constraints. Organizations must commit only what is really achievable under any given circumstances. The reputation of an organization increases if they under-promise and over-deliver, rather than over-promise and under-deliver.

- **Budgets:** Though it is impossible to get approvals for a luxurious budget, it is absolutely necessary to have all IT-related costs like hardware, software, telecom, installation, cables, support, upgrades, licences and other essential costs clear, with a reasonable buffer for each. In addition, other costs like taxes, transport, travel, insurance, people, fees, etc, must also be adequately covered to arrive at an overall budget. Every single bit of cost and effort should be accurately documented and a budget presented and approved. Provision must also be made for escalations in budget should the original requirements suddenly change for any reason.

- **Project deliverables:** The project manager of an IT project must have a clear and detailed picture of what is being achieved or delivered to a customer or the project sponsor. Commitments must be very clear and fully understood by both parties. Large projects have a particular tendency to change track for unexpected reasons, so it is always better to involve the customer at regular intervals to ensure that the project is on the path of achieving its objectives.

- **SLA:** It is always recommended to establish a clear SLA between the customer and the service provider to ensure that every aspect of the project is covered in writing. A typical SLA should cover the following main points:

- Name of the project or implementation.

- Number: a reference or a contract number.

- Description of the project.

- Client details.

- Parties to the agreement .

- Legal jurisdiction.

- Scope of work: detailed scope of work to be done by the service provider.

- Customer obligations: For example, providing electricity, computers, workspace, access to building, etc.

- Out of scope (both parties).

- Non-disclosure details.

- Obligations of both parties.

- Budget: who is going to pay for hardware, software, telephone bills, etc?

- Timeframes: start date, end date for the project.

- Assumptions.

- Constraints.

- Risks.

- Financials: budgets, taxes, payment terms, approvers, penalties, etc.

- Service windows: hours of work, days, holidays, etc.

- Communication methods: agreed methods of communication, eg, e-mail, fax, telephone, etc. For example, it may be mandated that all technical changes to the original specs must be communicated via e-mail and also via fax.

- After-hours support and costs.

- Staff involved and their roles.

- Managers and their roles.

- Training requirements.

- Incident-handling procedure.

- Problem-handling procedure.

- Change management procedure.

- Reports and metrics: what standard reports will be exchanged?

- Escalation procedure.

- Project termination clauses, notice period.

- Signatures.

- Appendix.

Service level agreements must be prepared by involving the following departments from both sides to avoid complications and misunderstandings later:

- Technical departments for understanding and defining the IT deliverables.

- Finance departments for defining the financial implications.

- Legal departments for understanding and minimizing legal complications should there be a legal battle later.

- Others: if required.

Critical projects are often started without establishing proper documentation, leading to all kinds of misunderstandings and confusions later. SLAs don't prevent project failures, but they do minimize the impact of a failure due to expectations mismatch, cost factors, post implementation hassles, etc.

- **Killing projects on time:** Many projects run into rough weather for various reasons. Killing a bad project may actually save reputation and further wasted costs. However, it is not an easy job to kill a project, and the people who do the actual project work can rarely take such decisions. Such decisions can only be taken at senior levels and the biggest problem is to convince them to do so. Secondly, there could be various official and political compulsions that prevent bad projects from being killed. However, it may always be in the best interests of an organization to recognize and kill bad projects on time before they cause further damage.

- **Learning from past mistakes:** Experienced project managers know that there are many things that can go wrong in spite of detailed planning. Some can be predicted, many cannot. However, it is important to learn from known mistakes and problems faced in the past. Project managers should cultivate the habit of learning from others' mistakes in order to avoid repeating the same mistakes themselves. For example, many organizations routinely over-commit on various aspects and end up with problems later. IT project failures are often devastating to an organization. Abnormal delays, bug-filled software, missing features, expectation mismatches, etc, can mean the end of a project, organizational reputation or even financial ruin for a company.

- **Resource backups:** Projects can fail due to resignations, accidents, health issues, etc, of key people. All these can adversely affect a project. A project manager must ensure that all risks related to key people, including the project manager himself, are adequately covered by a suitable backup plan. For example, resignation by a key member of staff is hard to prevent, but good knowledge management and role sharing would reduce the impact on the project.

CHAPTER 11: INFORMATION SECURITY

'When humans are too happy, even the gods are jealous.'

Old jungle saying

What is information security?

Organizations can suffer from various disasters if critical information and data is compromised by any means. Organizations will rely on several types of data. Some of the information contained in them can be confidential and must not be viewed or altered by unauthorized persons. For example, the salary details of all employees cannot be made public for everyone to know or view. Or a company payment website can be breached by hackers, causing reputation losses. Hence, it is necessary to have a protective envelope around the various kinds of data that an organization uses. This is information security. An organization should classify all its data appropriately and ensure proper safeguards for each. Some examples of classifying company data are:

- **Confidential:** For example, only the employee and certain types of employees like HR and finance should know an employee's salary: it should not be visible for others.

- **Secure or restricted:** Only authorized staff should handle passwords of mission-critical systems, production servers, etc.

- **Internal or private:** For example, company policies can be made visible to all employees via the company intranet.

- **Important:** A software development team can classify its software code as important and can restrict access only to certain team members.

- **General or public:** For example, certain information like fire safety, health tips, first aid, etc, can be classified as general.

Safeguards can be of several different types depending on the nature of an organization:

- All confidential data can be housed in a secure file server with access only by the authorized department's personnel. The administrative password can be kept in a secure safe and all usage logged in a register.

- All important data can be stored in secure file servers that can be accessed only by authorized employees.

- People can be prevented from printing, photocopying or e-mailing certain types of document. For example, it is possible to convert many types of document in an Adobe PDF file and have printing disabled to make it read-only.

- Ensuring that all hard copies of confidential documents are shredded after use so that they do not fall into the hands of unauthorized persons.

- Ensuring that nobody stores important data on laptops, CD-Roms, diskettes, etc, that can easily get misplaced or stolen.

- Electronic systems can be implemented to log and monitor activities on all computers or only by certain users.

... and other industry best practices or manufacturer's recommendations.

11: Information Security

What are the various ways in which information security can be compromised?

As mentioned earlier, disasters can happen to organizations if information security is compromised in any one of a number of different ways, eg:

- An organization connected to the Internet can be hacked by unauthorized persons if there is no proper firewall, intrusion detection or anti-virus system. A firewall will prevent an organization's internal IP network being visible to the outside world. An intrusion detection system will spot suspicious activities happening on the network. For example, an intrusion detection system can detect if some rogue software program is initiating a 'denial of service' attack on a website.

- A laptop containing confidential and sensitive information can be stolen.

- Confidential documents can be scanned or photocopied by unauthorized staff.

- Unauthorized personnel can intercept e-mail.

- Unauthorized persons may get access to data centres.

- Tapes, CD-Roms, diskettes, pen drives, etc, containing confidential data can fall into the wrong hands.

- External consultants, contractors, vendors, etc, working within an organization can view or access confidential data they are not supposed to see.

- Carelessness and human error in allowing unauthorized persons to get passwords, entry by strangers, etc.

- Critical passwords getting lost, stolen or changed by unauthorized persons.

- Somebody can hack a company website and alter or steal sensitive information. For example, if an organization sells products over the Internet, somebody can hack into the website and collect customer information like credit card numbers, e-mail IDs, etc.

- Employees who resign or get fired may destroy important data before they leave, or pass on sensitive information to outsiders.

Example

One of the companies that the author was working for in the Middle East had supplied several computers to a large defence organization. Security and movement of materials were extremely tight. The author's company was maintaining the hardware. However, whenever there was a hardware fault, like a hard disk failure, there would be two qualified defence personnel meticulously supervising the disk replacement by the IT vendor. Even after replacing the failed hard disk with a new one, they would not allow the vendor to take the failed hard disk back, irrespective of the nature of the fault, for security reasons.

What is hacking?

Hacking means gaining unauthorized access to a computer, its files and programs. The people who do this are called hackers. Hacking may happen just for fun or for commercial gain. An outsider can hack into an organization's network and somehow get access to critical or sensitive information. Sometimes hackers may destroy or copy important data.

11: Information Security

How can organizations prevent hacking?

Some of the common methods to prevent hacking are as follows:

- Install a state of the art firewall (hardware or software) between the company network and the Internet. Firewalls prevent a hacker sitting on the Internet from snooping into an organization's network. Excellent firewalls are available from companies like Checkpoint, Cisco and others.

- Install all manufacturer-recommended patches, hot fixes and service packs on all computers. These patches fix various vulnerabilities that can be exploited to hack into a machine.

- Change critical passwords often. Ensure that the passwords are not easy to guess. For example, don't name the passwords 'blank', 'password', 'secret', etc, that are easy to guess. Have a combination of alpha-numeric and uppercase/lowercase letters.

- Always have the latest anti-virus update on all critical systems.

- As an added precaution, purchase and install personal firewalls on computers. This software can detect and alert a user if some other computer is trying to access his or her computer. Personal firewall software helps prevent people from hacking into your computer while you are on the Internet. A personal firewall can help make your computer impenetrable to hackers. However, as with anti-virus programs, it is important that you keep your personal firewall software up-to-date.

- Install Spyware and Adware removers. Spyware, Adware, etc, are tiny programs that install themselves without your permission, while you are browsing the web. Anti-virus programs usually cannot detect such programs. Depending on how it is written, these Adware and Spyware can send out send out sensitive information without the user's knowledge. Visit *www.bulletproofsoft.com* and *www. download.com* to download or evaluate various types of adware and spyware removal tools.

- Do not open suspicious e-mail attachments. The attachments can contain a Trojan virus or a program designed to wreak havoc on your computer. A Trojan is a computer program which, when running on your computer, allows a hacker to gain easy access to your computer system. Users cannot see that a Trojan program is running on their computer. Some Trojans allow hackers to take control of your computer system.

- Don't download stuff recklessly from the Internet. Websites designed by troublemakers provide software which is intentionally designed to hack and damage your computer.

What is port scanning?

A computer operating system and networking software (TCPIP) has various software ports for various functions. For example, port number 80 is used for accessing the Internet, port number 25 or 21 is used for file transfer, and so on. There are 65,536 such ports. Not all ports may be used or active on a computer. A port scan is a series of messages sent by a hacker attempting to break into a computer to learn which computer network services the computer is vulnerable for. Port scanning is a favourite approach of computer hackers, giving the assailant an idea where to probe for weaknesses. Essentially, a port scan consists of sending a message to each port, one at a time. The kind of response received indicates whether the port is used and can therefore be probed for weakness. Closing ports can prevent port scan and intrusion. For example, if the computer is not used for file transfer, it should not have the FTP services enabled. It is not possible to check and close all the unwanted ports manually. Specialized tools are available instead, from websites like *www.lantools.com*, that can scan a computer and generate a report of vulnerable ports along with advice on how to close them.

CHAPTER 12: DISASTER RECOVERY TOOLS

'It is only when they go wrong that machines remind you how powerful they are.'
Clive James

Introduction

This chapter will outline various IT tools and services available to implement disaster recovery. The information is based on the respective websites and brochures: mention of these tools is not an endorsement, promotion or sales recommendation of the products or services. Organizations are advised to fully evaluate the features of the products to see whether they suit their business, technical and commercial requirements. Not all tools may be required for all organizations. Most of the websites allow you a time-bound evaluation copy for download.

Backup tools: A reliable data backup software is the most important and basic system requirement in any organization. Some of the high-end solutions that are available from reputable manufacturers like Veritas and Hewlett Packard are listed below. Visit their websites for detailed information and specifications.

- VERITAS Cluster Server is a high-availability solution. Cluster Server is ideal for reducing both planned and unplanned downtime, facilitating server consolidation, and effectively managing a range of applications in heterogeneous environments. Visit *www.veritas.com*.

- Enterprise backup / recovery / archive hardware. HP StorageWorks Enterprise Backup Solution (EBS) is a complete solution built around HP StorageWorks tape-automation products such as the HP StorageWorks ESL9000 Tape Libraries and the HP StorageWorks Ultrium 460 Tape Drive. Visit *www.hp.com*.

- Replication/data availability solution. In the disaster-recovery arena, VERITAS has a replication / data availability solution called VERITAS volume replicator.

Standard backup tools just allow backups onto a tape or other backup device. However, this may not be suitable for disaster recovery. Additional tools may be needed to ensure speedy disaster recovery and business continuity, eg:

File replication tools: These tools can be used to copy or mirror files from one disk to another, or from one server to another. Various options and schedules can be set on what, when and how to copy.

Example 1: File Replication Pro (*www.filereplicationpro.com*)

This is a tool that allows a user to move, copy and synchronize directories and files between a source and a destination by creating jobs. A job is a listing of the source path, destination path and the file operation you want to perform such as move, copy, one way sync or two way sync. It also has scheduling and schedules range from immediate, run once at a particular time, run on an interval or run continuously. Schedules can be quite complex and flexible.

Example 2: PeerSync tools (*www.peersoftware.com*)

This is a similar software program that has various file synchronization features. PeerSync Server runs directly from the server and synchronizes from desktop to server or server-to-server. It also allows for remote-to-remote synchronization.

12: Disaster Recovery Tools

Database synchronization tools: Regular file synchronizers are suitable only for regular files, documents, etc, not to synchronize databases as simple copies of the database elsewhere will not serve the purpose. Special methods and tools must be used to ensure the contents of a database and its dependencies are duplicated on another system in order to make it usable in a DR scenario.

Example1: CDB database tools (www.dbbalance.com)

This are tools that allow comparison and synchronization of heterogeneous databases. They are intended for use in an environment wherever there is a need for changes made to a local database to be pushed to a live database on a remote server. They support various flavours of database like Oracle, SQL and other popular databases.

Example 2: SQL tools (www.red-gate.com)

SQL tools have various utilities that take two databases, compareall the objects in each database and give the differences between them. In addition, they also allow automatic synchronization of data and objects from one database to another.

E-mail continuity tools: Nowadays businesses have become extremely dependent on e-mail. In many cases, businesses will come to a standstill if e-mail systems are down even for a couple of hours. However, it is not practical for organizations to have staff use two different e-mail systems and mail IDs to communicate. A few service providers provide e-mail continuity services for a fee, eg:

Example: Emergency mail system (*www.messageone.com*)

This is a standby e-mail system that is automatically synchronized with your primary e-mail environment. In the event of an emergency or planned outage, EMS can quickly be activated over the web or by phone. Once activated, selected e-mail users will have direct access to a fully functional web-based e-mail account that allows users to send and receive e-mail from their standard e-mail address. The standby account includes all of the key features of the primary e-mail system including contact lists, calendar appointments, distribution lists and important historical e-mail.

CHAPTER 13: INTRODUCTION TO NON-IT DISASTERS

'There are two big forces at work, external and internal. We have very little control over external forces such as tornadoes, earthquakes, floods, disasters, illness and pain. What really matters is the internal force. How do I respond to those disasters? Over that I have complete control.'   *Leo Buscaglia*

What are some of the non-IT disasters that could happen to an organization?

Within an organization every department will have its own importance and dependency on others. Every department must function collaboratively to ensure continued revenues and future business. Major equipment failures are not the only disasters that can happen. Disasters can happen in many other areas as well. Risks and potential disasters lurk everywhere. Some of the non-IT disasters that can strike an organization are:

- Trade or labour union problems.

- Project failures.

- Human error.

- Marketing and sales blunders.

… and numerous others.

This chapter outlines various non-IT disasters that can strike an organization. The aim of the chapter is not to cover every conceivable disaster comprehensively – entire books can and have been written on each type of disaster – but rather to raise your awareness as to the type of disaster that could befall your particular organization, to prompt you to analyse the specific risks you face and to encourage you to include the most relevant scenarios in your DR and BC planning.

What are the usual trade or labour union problems?

Many organizations and factories worldwide have strong trade or labour unions. These are necessary in many types of organization to keep a check on unfair labour practices, exploitation of workers by management, etc. But they can also cripple businesses at the drop of a hat, sometimes for trivial reasons. In some parts of the world disputes get out of hand and the union members resort to violence and damage to equipment. Organizations can go bankrupt if the labour unions do not listen to reason but equally the management should not simply ride rough-shod over their workers (if only because it rarely makes good business sense anyway). In many countries today there are strict laws governing what trade unions – and employers – can and cannot do. Despite such laws, however, strikes are still surprisingly common in the UK (50 days lost each year per 1,000 employees, according to the latest government figures) and even more so elsewhere. And strikes are simply the most extreme symptom of industrial relation problems. The following list of precautions an organization can take to minimise exposure to such problems is not exhaustive but will give you an idea as to the sorts of action you should consider taking, depending on the issues and practices in your part of the world:

- Have proper and transparent human resources (HR) polices for all classes of worker.

- Provide proper and reasonable employee benefits like medical facilities, housing, transport, etc.

- Have a proper profit sharing scheme among employees.

- Treat all employees with respect, dignity, etc.

- Follow all established, humane and legal procedures while hiring and firing employees.

- Provide employees with adequate financial support during critical illness, family troubles, etc.

- Have proper background checks to ensure that managers and supervisors treat employees and workers with respect. (There was a case recently of a lengthy union problem when a young shop floor manager abused an elderly worker in a factory.)

- Implement all guidelines and safeguards pertaining to the specific industry.

What is a project failure?

Nowadays, most companies – especially software companies – are involved in dozens of projects for customers. Projects can range from small one-member projects to project teams having hundreds of staff. An organization's ability to portray itself as a reputable player in the market depends on how well it can execute projects for its customers.

Projects can have dozens of risks, as outlined below. All these can derail projects and cause various types of disaster and loss. Badly executed projects have severe financial and legal implications, not to mention loss of reputation and decline in future business.

- **Inadequate budgets:** One of the biggest risks is not preparing an accurate budget for projects. Inadequate budgets have a corrosive effect on the entire project and will usually derail it.

- **Hiding bad news:** This is very common in many high-profile projects. In an attempt to impress senior management or customers, only good news and rosy pictures are presented. Bad news is hidden, delayed or diluted, resulting in a major crisis later. Hiding bad news can happen for a variety of political, strategic or improper planning reasons.

- **Hype and over-commitment:** Nowadays, hype and over-commitment are commonplace. Businesses and projects love and thrive on hype and flashy presentations to impress customers and kill competition. Companies may commit more than they can deliver and the project will start going haywire very quickly.

- **Technology issues:** Projects also have technology-related issues like improper sizing, selection of improper tools, etc, all resulting in serious business issues later. Technology is not the solution to all project ills. Excellent technology supported by poor documentation, project management or inadequately-trained support staff can still ruin projects.

- **Manufacturing defective and poor quality products:** In an attempt to beat competition and speed up deliveries companies often indulge in reckless cost cutting, thereby reducing the quality of the product. A poor quality product usually has an adverse effect on company revenues in terms of servicing, reputation and even legal issues when it comes to materials like safety equipment.

- **Poor project management:** It is quite common in many companies to have one person manage several complex technical projects. Such managers will usually be involved purely from a project management methodology view. They will have to depend on or trust others to ensure that things are going smoothly. As the project manager will not have in-depth subject knowledge from a real hands-on perspective, such project management can lead to disasters and project closures. For example, suppose there is a complex software development project. If a programmer claims or insists on three months timeframe to complete a critical piece of software the manager will not be in a position to judge whether that claim is realistic or whether he is being taken for a ride. Alternatively, due to insufficient knowledge the project manager may make impractical commitments to customers leading to severe problems later. In fact, poor planning and management, rather than technical issues, are the most common causes of project failure; the warning signs are:

  - Lack of user involvement: if the end-users aren't sufficiently involved and consulted at the outset, then the project is very unlikely to achieve its objective, however technically excellent it may be.

  - Unrealistic timescales: if too short a timescale, the project will either be skimped on to meet the deadline or it will be late, fatally undermining the business case for it and with serve knock-on effects for the rest of the organization, customers, reputation, etc; if too long a timescale, the project will drag and will turn into a 'white elephant', a solution to a problem that no longer exists.

- Inadequate specifications: if those charged with delivering the project are not given a clear enough brief, then the chances of a successful outcome are remote.

- 'Scope creep': this is where the project sponsor keeps changing his mind or adding extra requirements (usually because of a lack of user involvement and inadequate specifications in the first place).

- No change control system: just because the business requirement has changed doesn't mean that this can simply be mirrored in the project requirements. Better to stop the project and start all over again.

- Poor testing: the end product might withstand all testing the project team can throw at it but it's the users who count and if they don't run enough acceptance tests then, sooner or later, there is quite likely to be a very big problem (and the project staff may no longer be around to fix it).

• Other project-related issues like improper design, wrong technology, insufficient staff, etc.

There is no shortage of books and reading material as to how and why projects fail. The US-based Project Management Institute (the world's largest project management organization) is a good place to start: *www.pmi.org*.

What is a human error?

Human errors happen when people do something they are not supposed to do, either deliberately or by mistake. Myriads of human errors are possible. Human errors, from both management and staff, account for thousands of small to gigantic disasters. A techie may simply use a wrong computer command and wipe out years of data in seconds, or an electrician may accidentally cross some wires and blow up the entire electrical system. Human errors and their impact can be minimized by:

- Keeping things simple.

- Having only authorized and trained staff handle sensitive functions.

- Providing adequate training to all employees.

- Providing user education, 'do's and don'ts', etc.

- Following professional management practices. For example, a bold decision by the management to hastily dismiss a popular employee may result in abrupt union problems resulting in the factory being set on fire.

Every human error is a learning opportunity: make the most of it. Analyse it and take all necessary steps to prevent a recurrence not just of that particular error but of ones like it. Then be grateful that as a result of that particular employee's blunder you now have a more robust organization, so don't take action against them unless their action was malicious.

What are marketing and sales blunders?

This is a juicy one. To tackle today's business world of cut-throat competition, organizations resort to every legal (and some not so legal) trick to lure customers to their services and products. Ever increasing business pressures lead to marketing and sales chaps promising the moon to every existing or potential customer they meet, which later results in severe business issues.

13: Introduction to Non-IT Disasters

Example 1

There was a particular 'top' salesman in one organization the author was employed long ago. He could easily clinch tough orders that other sales chaps could not. But his secret of clinching orders was to usually promise (in writing) free, expensive software to customers buying the company's computers. When the equipment was delivered the free software would be missing resulting in severe verbal duels, delays in payment, non-payment, etc. One certain government organization did not release payment or return the computer till the agreed software worth more than the computer's value was supplied since the sales chap had committed everything in writing. In addition, the government organization also blacklisted the company from bidding for its future requirements of dozens of computers worth several thousand dollars. The sales chap later lost his job and the company lost its reputation.

Example 2

A bright young marketing chap working for a manufacturer of domestic appliances in the UK thought it would be a terrifically good idea to do a deal with an airline and then offer a free flight to New York for every appliance bought. He forgot that the airline ticket was usually much more expensive than the appliance, so everyone who was planning to go to New York anyway bought the appliance and demanded a free ticket. The company had to buy so many extra airline tickets (no longer at a good discount) that the whole exercise cost the company £40 million. The bright young marketing chap lost his job. So did his boss. And his boss's boss.

Thousands of marketing and sales chaps commit various such blunders throughout their careers by promising things the organization cannot deliver. The situation has not changed much even now. In fact, it is done in a more polished manner by showing fancy presentations about the wonderful services, products, return on investment, etc, that the customer would gain if they bought the products or services. Unbelievable SLAs are promised and even signed, with the supplier being fully aware that such commitments cannot be met. They might hope to rely on some small print in the contract, but these blunders eventually boomerang on the organization and result in several financial, reputation and legal risks to the business. More and more customers are taking companies to court due to false promises made in fancy and false advertisements.

What are financial disasters?

A financial disaster is anything that seriously and negatively affects the company's revenues. Any of the other disasters discussed in this book can have a negative financial impact, but one of the things that worries your CFO most is exposure to fraud (or the risk of being caught!). Fraud comes in many different guises:

- Employees cheating the company, perhaps through an invoicing scam using a partner outside the organization to help them.

- Suppliers cheating the company, perhaps through over-billing or supplying sub-specification materials.

- Customers cheating the company, eg, simply not paying.

- Top management cheating the company – and everybody else. The Enron scandal in the US led to the downfall not just of the company (with staff losing pensions as well as jobs) but also to the downfall of their accountants. This in turn led to the Sarbanes-Oxley Act, masses more expensive regulation for American companies, de-listings from the New York Stock Exchange, etc, etc.

13: Introduction to Non-IT Disasters

Example: Recent news report of a financial fraud

Former WorldCom chief executive Bernard Ebbers, accused of masterminding a massive accounting fraud that led the telecommunications giant into bankruptcy, was convicted on charges of fraud and conspiracy by a US court. The conviction came after five weeks of testimony and more than 40 hours of deliberation over eight days, with the main witness against Ebbers being Scott Sullivan, his trusted aide and former finance chief of WorldCom. Sullivan testified that Ebbers had ordered him to commit fraud by insisting that the company had to show performance and earnings up to the expectations of the Wall Street Journal.

Every organization must ensure that its financial systems, processes, policies, etc, are in proper order and within legal limits and government regulations.

What are some of the common recruitment risks?

Hiring anyone is a risk. This is not a new type of disaster. It is as old as the industrial revolution. At one time, management would only hire highly-learned people who knew their stuff, or would hire trainees and train them to the fullest possible extent. For example, until about a decade ago it was not possible for anyone to be promoted to a manager position until they had slogged and understood the trade properly, as well as the tricks of the trade. Nowadays, it is not always necessary to have any real experience to become a senior manager in charge of a complex project. A young inexperienced, but flashy candidate can easily become a senior manager or even a chief executive officer of a firm, and start bossing around experienced staff, leading to loss of good and experienced staff. Astronomical salaries are commonplace. Hiring the wrong people can cause various types of disaster. Some of the common ones are listed below.

- A bad mannered or unprofessional manager or supervisor can cause a severe degradation of employee morale and customer satisfaction.

- Good and experienced employees may resign due to bad supervisors causing project failures or delays.

- Employees who don't treat customers right can cause business losses or even lawsuits.

- Employee harassment and workplace bullying can lead to reputation losses or hefty lawsuits.

All these can be described as recruitment disasters, which eventually affect the business. One particular new organization (currently bankrupt) used to pay nearly four to six times the industry average salaries to its entire senior staff. The company later wound up severely in debt and lost its reputation within a year as the business managers did not deliver anything. The entire dot com bust was mainly due to inexperienced, under-age chaps becoming CEOs of their own organizations without having a clue about running a sustainable business.

What is a natural disaster?

A natural disaster is something like a fire, flood, earthquake, weather disturbance, etc, that cannot be controlled by mankind. However, some precautions can be taken to ensure that the losses caused by these natural disasters are minimized. For example, nowadays many builders are not building strong and rugged structures like they used to build decades ago. Modern buildings are full of glass, wood, carpets, curtains, fancy materials, etc, that can catch fire or get damaged easily. All these things increase business risks dramatically. Hence, businesses now have extra things to worry about, calculate and mitigate risks for. Except for fire, organizations and business managers cannot prevent other natural disasters like earthquake, floods, etc. The risk of fire can be greatly reduced by implementing every fire safety precaution known to man. Some of the common ones are listed below.

- Have plenty of fire extinguishers around and test whether they are working frequently.

- Train as many employees as possible to operate a fire extinguisher.

- For areas where it is possible, have automatic or manual water sprinklers installed.

- Always have overhead tanks and fire hydrants in working condition.

- Ensure that nobody parks their vehicles or loads unwanted material near fire extinguishers, fire hydrants, water supply, etc.

- Have the building and offices inspected by the fire department and other qualified consultants regularly. Implement all their recommendations.

- Have enough fire exits all around and keep their pathways free.

- Ban smoking in the organization. Don't allow fireworks during celebrations.

- Have electrical equipment tested and certified frequently.

- Inspect and remove all inflammable material throughout the organization.

- Avoid having flammable carpets, curtains, etc, if possible.

- Have the fire alarms in working condition.

- Practice fire drill and evacuation procedures regularly.

... and many other fire prevention mechanisms, as recommended by fire departments.

What about health and biological threats to employees within organizations?

This is a serious concern among several organizations and must be handled carefully. Here we are not referring to a terrorist or biological weapons attack on an organization, but biological attacks that happen due to unhygienic conditions or improper care. Organizations must ensure that the company water supply, air-conditioners, etc, are subject to periodic checks to ensure that there are no health related risks. If the drinking water supply tank in the organization gets polluted or infected with some bacteria or virus, the entire staff can become seriously sick within a day. If the cooling towers or evaporative condensers are harbouring bacteria then people can catch the fatal Legionnaire's disease. For an extreme example, a deadly accidental gas leak in the Union Carbide factory in India killed and maimed thousands of people in the factory and its surroundings in the 1980s.

Example

Manager: 'Why is everyone looking sick and pale suddenly? Is there some epidemic going on?'

13: Introduction to Non-IT Disasters

Admin Manager: 'We have no idea. We're calling a doctor to treat us.'

Doctor: 'The symptoms look like food poisoning. What did the employees eat or drink today?'

Admin Manager: 'Nothing unusual, I suppose. Everyone here eats in the lunch hall.'

Doctor: 'I am sending everyone home after treatment, and also ordering a thorough inspection of everything in the kitchen and other areas, including the water supply.'

Later,

Admin Manager: 'You were right doctor. We found a dead lizard in the drinking water tank.'

Some of the common methods to keep the organization safe from such threats are listed below.

- Check the water supply periodically to ensure clean filtered water is supplied everywhere.

- Have the ventilation clean, and gas cylinders, etc, properly secured and away from common areas.

- Keep all hazardous materials and chemicals in recommended containers.

- Ensure that the building security staff are trained in all first aid procedures.

- Have phone numbers of doctors, hospitals, ambulances, etc, handy and updated.

- Organizations that manufacture chemicals, explosives, medicines, etc, must have very, very strict safety guidelines to ensure their own employees as well as the entire surroundings are protected from harmful effects.

- Ensure compliance with all health and safety regulations and legislation. If you're not sure what regulations apply to your organization – find out. (In the UK the best starting point is the Health and Safety Executive: *www.hse.gov.uk*).

What about electrical failures and blackouts?

No organization can do its business without using electricity. Electricity is the lifeblood of many organizations. Electrical faults and blackouts account for a large percentage of disasters worldwide. A simple short-circuit in a factory can set the whole factory ablaze. Lengthy power failures or blackouts also affect business. Electricity can kill and maim people if not handled with care, so organizations must ensure that their electrical systems are always in perfect working condition. Some of the common methods to prevent electrical disasters are listed below.

- Have the building wired only by professional electricians.

- Ensure that all electrical equipment has the necessary fuses, overload trippers, safety handles, insulations, etc.

- Do not have any loose wires hanging around anywhere.

- Ensure that electrical points and wiring blocks are safely shielded from unauthorized personnel.

- Ensure that there is no water seepage near electrical items.

- Ensure that the cables and connectors used are of high quality even though they may be expensive. Loose connections and faulty connectors are the number one cause of electrical failures.

- Have qualified electricians inspect and certify the building wiring systems.

- Ensure that pests, eg, rats do not cut wires.

- When equipment is not in use do not keep it powered on. Switch off all unnecessary electrical gadgets when leaving the office.

- Have proper UPS and generators for critical equipment.

- Always draw the recommended amount of power. Never short-circuit fuses just to draw more power.

- Have torches, batteries and emergency lights at all required places.

… and many other recommended procedures. Use a certified electrical consultant for best results.

What precautions can organizations take to handle civil disturbances?

Nowadays, civil disturbances are quite common in many countries. Unemployment, religious attacks, over-population, terrorism, religious processions, political causes, etc, all lead to civil disturbances. There are millions of youths around the world without jobs loitering on the streets hoping for some free action. Civil disturbances can start and blow out of proportion very quickly. Agitated crowds can easily set fire to a building or vehicles, attack innocent people, etc. The frenzy of an agitated crowd cannot be controlled easily.

13: Introduction to Non-IT Disasters

Example

Several years ago, a particular newspaper had written a nasty article hurting the religious sentiments of a particular community. Very soon enraged mobs stormed the newspaper office and set fire to the whole building. The agitation soon spread out of control, and surrounding shops and business establishments were also burned down. Finally the police had to resort to firing on the mob to disperse and control the situation.

Organizations can ensure some degree of control over, and protection from, civil disturbances by following these measures:

- Agitated mobs love stones, sticks, fire, etc. Hence, ensure that the organization's surroundings do not have easy access to such material for a mob's consumption.

- At the first sign of a civil disturbance, ensure you call the police and fire department quickly to protect the building, employees and other property.

- If possible, send the entire staff home or to safe locations using all available means of transport to prevent injuries to employees if a riot starts suddenly.

- Tell employees not to try any heroics or unnecessary engagement with the rioters.

- Follow all recommendations of the police and fire departments.

- In case a riot still happens, ensure if you can that no employee is harmed. Transport injured employees to the nearest hospitals as fast as possible.

How can organizations take precautions against terrorism?

Nowadays, terrorism is spreading worldwide at an alarming rate and governments and military forces are finding it extremely difficult to control it. A few, but not guaranteed, methods of avoiding terrorism are listed below.

- Ensure that the organization does not have main branches, large offices, etc, in states and countries prone to terrorism.

- Assess countries and geographical locations by order of safety in terms of crime, terrorism, political turmoil, religious preferences, etc, and ensure only the minimum amount of staff and property are exposed to such risks.

- Constantly assess risks and political scenarios to evacuate staff if necessary. There are reputable consultancies who specialise in this sort of analysis, eg Control Risks (*www.crg.com*).

- Ensure, if possible, that employees do not take their families to political and other religious hotspots.

- Hire and train locals as employees.

- Keep good relations with local politicians if possible. This is really helpful in many third world countries where political parties wield enormous power over the business community.

- In areas where terrorism and crime are rampant, try to indulge in charitable works and offer more employment for local people. Organizations must develop a social and community approach to generating more jobs. More and more employment means less and less crime, trouble and other disturbances on the streets. For example, a reputable Indian industry situated in the northeast is providing free hospitals, ambulances, schools and other free social service to thousands of unemployed people.

- Have highly trained security guards and protection forces wherever possible.

What is a travel-related risk?

Until a few years ago, businessmen considered business travel a prestige and luxury. Travel by plane, staying in five star hotels, etc, was considered a status symbol. It still retains some of its past glory, but travel nowadays is considered a big risk. The increase in terrorist threats, civil disturbances, airport security, global unrest, political instability, etc, worldwide is creating a lot of travel-related risk. It is not possible to travel peacefully on any business trip without considering the numerous risks associated with it. Terrorists have blown up planes, airports, train stations, etc, for various political and other causes, killing and maiming hundreds of passengers. Today, businessmen prefer to avoid travel, unless it is absolutely essential. International travellers face a variety of threats that can disrupt a business trip, along with personal risk. Getting stranded in an alien country with little knowledge of local issues or escape routes can be devastating to any businessman. In view of the risks, organizations and governments are classifying countries and states into various categories that will be helpful in deciding whether to go on a business trip or not. Some of the common classifications are listed overleaf.

Figure 5: Travel-related risks

Country	Low risk	Medium risk	High risk
A	–	Crime, violence and theft	–
B	–	–	Terrorism, bombs, continuous conflict, war
C	Avoid crowded places; poor medical treatment	–	–

Nowadays, many insurance organizations offer insurance and evacuation services for certain sensitive countries. In the event of an emergency, the insurance holder can dial a toll-free number and get immediate assistance, cash, airlifting, etc, in the foreign country. Risk assessment teams are being formed in various organizations whose main business is to constantly assess various types of travel and other risks to the organization.

Example

A couple of years back, a certain software development firm lost several of its qualified IT staff due to a plane crash while they were travelling together on project work.

Some of the common methods to minimize travel-related risks are listed below.

- Avoid business travel in groups. Split the group and take different routes or flights.

- Use video-conferencing wherever possible, particularly to avoid short trips.

- Use a reputable travel agent and get the complete itinerary down to the minutest detail.

- Study the political conditions of the destination countries before travelling.

- Follow all rules and regulations of the host country.

- Keep all medicines and emergency numbers handy. Always carry a reputable mobile phone with international roaming facility enabled. Remember to carry its charger and a multi-socket connector for different countries.

- Memorize phone numbers, addresses, and e-mail IDs if possible.

- Don't travel without identification papers and local contact numbers.

- Always arrange for some reliable local help through the travel agent. Many foreigners get robbed, overcharged or harassed in various countries by locals.

- Avoid travelling independently in foreign countries. Always take some trustworthy local help if possible.

For further information on business travel hazards see www.rothstein.com and/or consult your country's embassy in your destination country.

What about the psychological effects of a disaster on employees?

Depending on its nature, a disaster can have a severe psychological impact on employees. During a crisis situation, personnel and family matters take priority over resuming business. For example, if there is a fire and many employees are hurt or badly burnt, it will not be possible for them to ignore their personal sufferings and start concentrating on disaster recovery or business continuity. Organizations must be prepared to handle the psychological factors associated with a disaster. They should first ensure that employees' welfare and safety take priority over instant disaster recovery or worries about loss in productivity. Disaster recovery and business continuity must be as humane as possible. Also, every effort must be made to ensure that employees continue to receive their wages or acceptable salaries in the event of a major disaster.

What is a reputation risk?

In today's highly competitive business environment, an organization's reputation in the eyes of the customer or stakeholders is extremely important. Reputation sells the organization's products and services. Reputational loss can harm an organization in very severe ways. For example, once famous and powerful organizations like Arthur Andersen, Enron, etc, collapsed because of reputation issues. Often, it is not necessary for an organization to commit a deadly crime for its reputation to take a plunge. A single case of harassment of an employee by one of its managers, if leaked to the press, can cause irreparable damage to an organization. A false and unproven allegation in the newspapers by some naughty reporter can also cause severe damage. For example, if a car owner lodges a legal case against a car manufacturer alleging that the brakes are faulty and this attracts publicity, then business can drop drastically even though it could be a one-off case of a brake failure in a single car. Organizations must guard their reputation very, very carefully in order to remain in business. Even the best PR managers cannot rectify a ruined reputation. Some of the common methods to safeguard an organization's reputation are listed below.

- Employees must be trained and informed not to blabber anything and everything about the organization to outsiders. For example, no employee must be allowed to speak to the press or journalists regarding any company issues. It should only be handled by the authorized press representative or the crisis management team.

- Organizations must ensure that their commitments to customers are met in accordance with their promises.

- Organizations must study customer complaints and handle them appropriately and in a timely manner.

- Organizations must ensure that unauthorized persons or other organizations do not use their names, trademarks, logos, etc, directly or indirectly, without permission.

- Organizations must ensure that they follow all professional, legal and other formalities. For example, adequate safeguards must be in place to ensure that there is no employee harassment of any sort in any department.

- Incidents that have the potential to blow out of proportion must be handled speedily and effectively. For example, if there is an unacceptable rumour floating around about the company's products or services then the company should immediately call a press conference, or put a notice in the newspapers explaining the company's stand.

- Organizations should have proper non-disclosure agreements with vendors, suppliers, consultants, etc, with strict penalties to prevent inside information from leaking out. For example, an external HR consultant hired by Company A to improve sagging employee morale should not go to Company B and talk about the poor levels of employee morale in Company A.

If your organization is too small to have a PR department, or if your existing PR staff or consultant do not have experience of crisis management, then you should identify a PR consultant who specialises in crisis management. Get them to prepare a crisis management communication plan for you and agree with them under what circumstances, and at what cost, they would be able to act for you in the event of a crisis. If the chairman's screaming that the company is in danger of imminent collapse unless something can be done about the press reports, that's not the time to start thinking about finding a PR consultant and negotiating a fee with them - you want them in the office within the hour, helping sort out the crisis.

What about industrial espionage?

Industrial espionage is the theft, spying or sabotage of an organization's confidential information by competitors, enemy countries, spies, etc. Industrial espionage isn't just in James Bond movies: it is all around, real and deadly. For example, if a car manufacturer is secretly stealing designs of new car models from another manufacturer, it is industrial espionage. Industrial espionage is very difficult to control nowadays with the easy availability of e-mail, tiny storage devices, etc. A rogue employee can easily steal many confidential documents, expensive designs, etc, from his organization and sell it to a competitor. Mobile phones and easy Internet access give opportunities to leak out company information within seconds. Or somebody could steal a laptop containing sensitive information and leak the company secrets all over the place. Employees (who may be secretly working for personal gain, competitors, etc) commit a large proportion of espionage activities. A reputable organization can easily become bankrupt if all its product designs, research information, etc, are stolen and manufactured by a competitor. For example, a pharmaceutical company could have spent millions of dollars to invent a new kind of medicine, but a competitor can easily steal the final formula and release the product in the market and create a financial disaster for the original company.

Some methods to try to prevent industrial espionage are:

- Run proper background checks whenever you hire new employees.

- Conduct periodic security inspections of essential processes. For example, the R&D department must ensure that they secure all their research work properly, and do not leave paper and important information lying around.

- Conduct surprise security checks on employees working in sensitive organizations. One defence organization conducts sudden body searches for all staff at random intervals to find out if any material or paperwork is being smuggled out.

- Cut down the number of laptops used, wherever possible, as they can easily get lost or stolen with sensitive information still inside.

- Ensure that all data is stored only on proper file servers and log all activities like copying of files to floppies, other drives, other computers, etc. Advanced logging and alerting tools are available to check illegal copying of files. For example, certain software tools do not allow files to be copied from their source location (maybe a file server) to any other location (eg, a C drive, floppies, e-mail attachments, etc) without a series of passwords, justification, approvals, etc.

- Do not allow consultants, contractors and other third parties to access sensitive data.

- Shred all paper documents before discarding.

- Use encryption tools to encrypt data.

- Do not allow floppies, USB disks, and other storage items to be used by regular employees.

- Ensure that sensitive information is not discussed in open areas.

- Have hidden 24x7 video surveillance in restricted areas.

- When an employee resigns, ensure that all login IDs, passwords, access controls, etc, of that employee are disabled or deleted so that they cannot be passed on to other people.

- Make sure your contracts of employment have stringent restrictions on use of confidential information, data, etc, if the employee leaves the company for any reason.

- Have a proper information security policy and keep monitoring the gaps, loopholes, etc.

- Install telephone-tapping devices if necessary, following proper legal guidelines.

- Prevent camera phones inside sensitive areas. Nowadays camera mobile phones are easily available and someone can easily photograph some sensitive stuff and send it out as an e-mail via the mobile phone without going through the company's computer network.

How can organizations prevent disasters relating to paper documents?

On a par with computer data, paper documents are the most important pieces of information that every organization will have and need to protect. Paperless offices are still far, far away. Most organizations still depend on paper for transactions, approvals, signature verifications, forms, etc. It is currently not possible to do away with paper. Organizations must store paper records for years and years, maybe forever in certain cases, for regulatory reasons. Even if the entire technical infrastructure is working fine, but the company loses all its paper records it is still a major, non-recoverable, catastrophic disaster, so elaborate precautions must be taken to ensure that all necessary paperwork is not destroyed due to fire, water or other risks. Some of the more common methods of storing information related to paperwork are:

- Install an electronic document management system. Scan every important document and store them as a retrievable electronic file, eg, a tiff or pdf.

- Keep all paper records in strong fire- and water-proof storage.

- Reduce and simplify the number of paper forms, documents, etc, wherever possible.

- Do a periodic cleanup and eliminate all unwanted paper. Have a quarterly or half yearly paper clean up exercise throughout the organization – but make sure first that everyone knows what they can't throw away (eg, tax records).

- Do not store any paper records near electrical systems, hot areas, humid areas, etc.

- Ensure bugs, mice, termites, cockroaches, etc, do not destroy paper records.

- If necessary, laminate key documents so that they are not damaged by moisture and water.

What other precautions can organizations take?

This chapter has introduced some of the more common types of non-IT disaster that might befall an organization, but there are countless others, many of which will be very specific to the particular organization, the sector in which it operates, its geographical location, etc. To adopt Donald Rumsfeld's simile, there are 'known knowns' which could affect any business (which your organization's regular systems and procedures should be able to deal with), 'known unknowns' (possibilities for which you have to be prepared, events which might happen but whether they are disastrous for your organization or not may depend on your response to them) and 'unknown unknowns' (events which could not possibly have been anticipated). By definition, there's not much you can do about the last category, other than respond appropriately when it happens, but there *is* more that you can do about the intermediate category, the 'known unknowns'. One of the best methods is by incorporating rigorous PEST analysis in your annual planning exercise. PEST is a way of looking at the Political, Economic, Social and Technological factors that might affect your organization in one way or another.

- **Political** factors might include a potential change of government (eg, if your business is heavily-dependent on government contracts, what happens if the other party gets into power) or a change of law (eg, online gambling businesses with a preponderance of American customers got a nasty shock when the US finally outlawed online gambling and arrested a couple of CEOs).

- **Economic** analysis can help challenge your basic assumptions about your market, both domestic and international.

- **Social** factors might affect the demand for your products and services or your ability to provide them (eg, there might be a backlash against advertising certain types of product – cigarettes, fur, fatty snacks for children, etc – which might affect you if you manufacture those products or supply companies which do; or perhaps you are finding it harder to get the staff you need because the local workforce isn't sufficiently skilled or they can't afford to live where you are located).

- **Technological** developments might also affect the demand for your products and services or your ability to provide them (eg, can your personal service be replaced by an automated competitor? is your product about to become superseded by a more advanced rival? are you relying on out-of-date technology?).

The unthinkable does happen, organizations do collapse, firms go bust. If you want to avoid that level of disaster, then it is worth making a point of conducting PEST-type analysis on a regular basis – preferably as part of the annual business planning cycle – to ensure that you don't become so immersed in keeping on top of the minutiae of running the business that you fail to anticipate the key event that destroys your business.

CHAPTER 14: DISASTER RECOVERY AT HOME

'Be grateful for the home you have, knowing that at this moment, all you have is all you need.' *Sarah Ban Breathnach*

As home PCs and Internet access have become more and more widespread, so the number of people working from home has increased enormously over the last few years. Whatever the size of your organization, it is almost inevitable that you have a number of people working from home, whether on a regular or occasional basis, from the CEO burning the midnight oil on the annual business plan, to sales people working up their Powerpoint presentations for a pitch the next morning to data-input teleworkers. In their own way, these people all expose the organization to risk of one sort or another. Any DR or BC plan must take account of this. This chapter outlines some of the risks to be taken into consideration.

What are the main risks associated with home working?

Working at home brings with it various major and minor risks not usually associated with working in the office. Some of the key risks are:

- Children and pets: they can cause problems if they can get access to business material. For example, the famous scientist Thomas Edison lost hundreds of his research papers due to a fire accidentally set by his pet dog.
- Fire

- Burglary and theft
- Electrical short circuits
- Power outages

... and other home-related disasters.

What are some of the ways to prevent disasters occurring in homes?

Whether you work at home yourself or whether you manage those who do, you need to take precautions. The following list is not exhaustive, so use it not just as a checklist but also as a prompt to help you analyse the risks and identify the appropriate measures to take.

IT-related precautions

- Do not load games and other fancy stuff on your business computer at home.
- Buy a reliable tape drive or backup device and back up your data regularly. More important, learn to restore and verify your backups.
- Learn how to configure your Internet connection, install software, anti-virus programs, etc on your own.
- Have a print-out of all important phone numbers, e-mail ids, vendor contacts, etc, and keep it updated periodically.
- Have an anti-virus and firewall system to protect your computers and keep it updated. Home-based firewalls can be purchased for less than US$100.

15: Plenty of Questions

- Think about security. A survey in the UK claimed that hackers are gaining access to corporate networks by exploiting lax security on over 350,000 home workers' PCs connected to their work IT systems. According to the study, one in six PCs tested were completely without protection. With well over two million UK employees using their home PC to access work networks, it is possible that 350,000 employee PCs are acting as back doors for hackers to attack business networks.

- Have your computers, printers, UPS, etc, under proper hardware maintenance contracts.

- Download your business e-mail to your computer, but also retain a copy on the ISP server.

- Have two e-mail IDs if possible, and configure e-mail ID1 to send a one-way copy of all e-mails to e-mail ID2 for backup purposes.

- Do not open attachments and other suspicious e-mails, which do not seem to be business related.

- Scan important documents and store the images on a CD-Rom or disk.

Safety-related precautions

- Throughout the UK and Europe, and probably in numerous other countries, employers have legal responsibilities in relation to health and safety and other issues affecting any of their employees who work from home. In the UK, the Health and Safety Executive have published a free 12-page booklet on the subject; you can download it from: *www.hse.gov.uk*.

- When working from home, use a separate room if possible. Keep all work documents, computers, diskettes, CD-Roms, phones, etc in a room that can be locked.

- Ensure that the business room is fire-proof, water-proof, pest-proof and child-proof. It should also be clean and tidy.

- Do not share your business computer with children, friends, relatives, etc – have a separate computer for them to use instead.

- Ensure that important documents, and other business related materials, are out of reach of children.

- Have a small fire extinguisher handy.

- Have a UPS with adequate power backup.

- Ensure that your electrical outlets are safe and properly earthed.

- Review your insurance policies. Does your work insurance cover losses at home? Does working at home invalidate your personal and home insurance?

- Do not leave laptops and other important business material in your car. If the car gets stolen, or mowed down by a truck, you will lose important data and you will probably not be able to claim under your insurance policies either.

- Take any other safety precautions necessary depending on the unique nature of your work, your home, location, availability of support, etc.

- Your home probably wasn't designed to be a work environment too, so be careful about overloading the electricity supply with too many plugs and extension leads going into too few sockets. Avoid having leads trailing all over the place that you (or small children) can trip over. And try to ensure that your desk, table, workstation or whatever meets the basic health and safety advice: ergonomists recommend that you sit upright, not too close to the computer screen, with your elbows, hips, knees and ankles all at 90° angles.

Document and data management

If you work from home on a regular basis, decide how you are going to manage your documents and data. How do you take electronic files home – on disc, by email, on a memory stick? And how do you take the updated versions back to the office – in the same way? It's probably best to use a couple of methods. Email is fine for short documents, but if you're doing a five-year business plan it's a pain to have to email large Excel spreadsheets which may be too big for your home e-mail system. Likewise, there's nothing worse than staying up half the night to finish something, e-mailing it to your office and then getting in the next morning to find that the e-mail hasn't worked for some reason. Memory sticks are an ideal solution to this sort of problem, but they can be a security risk themselves if you lose them.

Whatever solution you arrive at, you also need to think about how you update and synchronize documents that you use both at home and at work.

Data backup for stand-alone systems

If your master files are all in the office, perhaps it doesn't matter to you if you lose whatever you do at home.

However, simple backup devices and methodologies are available to back up and restore individual computers. Some of the easy, tested and proven practices to ensure data backups are as follows.

- **Image backups:** A computer can be fitted with two hard disks. Disk C: can be used for loading all required software, data, etc. This will be the primary business disk and the second hard disk can be used as a backup disk. The entire C: can be taken as an image file on to the D: drive. Special backup utilities like Norton's Ghost can be used to transfer a snapshot of an entire hard disk on to another hard disk. If the primary disk fails then simply restoring the image can restore the computer back to its original condition. For example, suppose an image has been taken on 1 September at 3 pm. And suppose the primary disk crashes on 5 September. By restoring the image, the computer can be restored to the condition that it was in on 1 September at 3 pm. Depending on the periodicity of the backup, systems can be restored to the last available image state.

Figure 6: Image backups

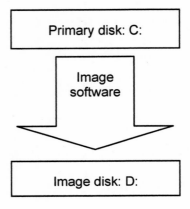

- **DVD backup:** As an additional precaution, the entire image file or essential files can be backed up onto DVD or CD-Rom disks. Low-cost DVD writers are available for less than US$150 (or may even be installed on your machine already, as part of the start-up package).

- **USB disks:** If only a few essential files have to be backed up then low cost USB disks ranging from 128 megabytes to 2 gigabytes are available in the cost range of US$50 to US$200. These disks are portable and easy to use but they will not help you to restore an entire computer.

- **Tape drives:** Low cost tape drives are also available that can back up data ranging from 20 to 80 gigabytes. Such drives and accessories cost in the range of US$300 to US$500, depending on features and quality.

Sample recommended solution

For large amounts of data backup it is better to have a combination of an image backup and a tape backup for complete safety. This will require three hard disks on the computer, if possible.

- The C: drive can be for the operating system, necessary applications, backup software, desktop settings, etc. This disk should not be used to store any important data.

- The D: drive can be used for taking an image backup of the C: drive. This disk should be reserved for the image file only.

- The E: drive can be used to store all data. This drive can be backed up into tape. If the tape drive has enough capacity it can also take a backup of the image file, for additional precautions.

15: Plenty of Questions

In the event of a disk crash the following methods can be used to restore the data.

- If the C disk crashes, a new C disk can be plugged in and the last image file can be restored from D: to C. This will normally require a bootable floppy or a bootable CD-Rom.
- If the D disk crashes, a new D disk can be plugged in and a new image backup of the C: drive can be backed up to the new D drive.
- If the E disk crashes, a new E disk can be plugged in and the data restored from the tape drive. This is possible because the tape software will be available on the C drive.

- **Laptops:** The above methods cannot be used on laptops, as it is not possible to have multiple hard disks on most laptops. Backing up laptops can be tricky as it is difficult to plug in tape drives and image disks. Some of the recommended methods are as follows.

  - USB disks can be used to take essential backups.
  - Use an external tape drive and REV drives like Iomega (_www.iomega.com_) that plug into the USB port for backup. Such devices have the option of creating a special bootable disk for creating an image backup of the entire hard disk. The user just boots from the floppy disk that allows DOS access to the tape drive. Next, an image utility like Norton's Ghost can be used to take a snapshot of the entire hard disk. A regular and periodic image backup will ensure that the system is adequately protected.
  - However, if a laptop disk crashes restoring it to its original state will involve some hassles and may require help from external vendors specializing in data recovery jobs. Such vendors have specialized devices that can extract data from failed hard disks, though they may not always be successful.

CHAPTER 15: PLENTY OF QUESTIONS

'How can we have wondered about so much for so long, and received so few answers?'
The Judybats

Here is a long assorted list of general and specific questions business owners can ask themselves, or the persons responsible for disaster prevention and recovery.

The most important question

This chapter contains dozens of questions related to DR and BC planning but there is one important question that must be answered before any work can start on creating a proper DR or BC setup:

How do you get commitment by top management for DR and BC planning?

It is easy to get the best possible plans, technical equipment, manpower, external consultants, etc, for establishing a proper disaster recovery setup if an organization is willing to invest the right amount of money. However, none of this is achievable if senior management, the decision-makers in an organization, are unable or unwilling to spend money on it. The inertia could be caused by financial or political factors or simply lack of knowledge.

An organization may have hired some IT staff or an external vendor to provide tech support for an important server. But, speaking from a business perspective, the IT staff, operators or external vendors are not really the owners of DR or BC for the organization. For example, if the server blows to pieces the IT staff cannot be held responsible for the organization being unable to conduct its business. Actually, the true owners of DR and BC are the business managers of an organization. They should know or understand what is the loss in financial, reputation or legal terms due to stoppage of various critical businesses and IT functions. They are responsible for ensuring provision of necessary , budgets, manpower, resources, alternative methods, etc, to tackle and prevent disasters.

Hence, for a successful DR or BC setup, the drive for it has to come from the top. It is usually the job of a CTO (Chief Technical Officer) or CIO (Chief Information Officer) to prepare a convincing DR or BC proposal with the necessary justifications to establish a DR or BC site. However, senior management and business managers are usually not interested in the technical 'nitty-gritties' of a DR or BC plan. They will mainly want to see a financial figure and why that amount has to be spent. A DR proposal therefore needs to cover all the following points:

- Aim and objectives. For example, why the organization must have a DR-BC plan facility.

- A list of all business critical functions and their priorities or importance. For example, our organization has the critical systems A, B and C.

- An estimated business or revenue loss if those critical functions stop beyond an acceptable timeframe. For example, US$5,000 loss per day if the main sales system fails.

- Reputation losses from disasters. For example, a newspaper report about the lack of DR and BC plan facilities for a reputable organization leading to the company stock nose-diving.

• Audit and regulatory requirements and penalties by not having a proper DR setup. For example, any penalties or fines for not sending some reports on time.

• A high-level plan of business alternatives available or planned for each of the critical functions. For example, whether business alternatives are possible manually or available for rent, hire, lease, and at what cost.

• A financial figure or budget of one-time and ongoing expenses involving hardware, software, manpower, non-technical stuff, real estate, and other expenses.

• Timeframes to establish.

• Some high-level details and diagrams on how it works.

• Other issues and requirements.

• Signatures.

If the above proposal is brainstormed and approved, the second-level work of in-depth technical and non-technical details can start. The key job of a CTO or a CIO is not only to write a convincing proposal, but also to get it approved and signed off and to start the activities. And after establishing a DR or BC site he or she must be able to prove that the proposed site is adequate to meet the current and agreed business requirements.

Questions on planning, security, etc

(Yes, No or N/A for each)

• Are our existing disaster recovery processes adequate?

- Are our offices close to airports and military areas prone to various threats?

- Are our offices close to factories and chemical plants that manufacture hazardous substances?

- Are there proper access control systems to prevent unauthorized systems entering the premises?

- Are employees allowing tailgating into the premises?

- Do we have proper security policy and guidelines published?

- Are our offices and workplaces fire-proof and water-proof? If not, what precautions do we need to take?

- Is physical security of our offices and workplace adequate?

- Are we sure that no unauthorized persons are entering the premises after office hours or during office hours, weekends or holidays?

- Are we sure that employees and other personnel are not passing on sensitive information to unauthorized destinations?

- Are our information security and classification adequate?

- Are we sure that sensitive information is not stored in unprotected laptops and local hard disks of employee PCs?

Questions on technology

(Yes, No or N/A for each)

- Are we sure that we are backing up all important data?

- Have we insured and properly labelled all equipment?
- Are we sure that laptops do not contain sensitive information?
- Have we ever tested restoring important data?
- Is there any obsolete outdated equipment or software we are still using which is not supported by vendors?
- Do we have sufficient redundancy on our telephone and other communication links?
- Are we following software-licensing guidelines properly?
- Is access to the data centre secure and to authorized persons only?
- Have we ensured that there is no electrical overload anywhere?
- Are all critical and sensitive passwords secured?
- Are we sure that no unauthorized persons are accessing our network?
- Is our website safe from hackers?
- Are employees writing their passwords on whiteboards provided to them?
- Do we have a proper firewall between our internal and external networks?
- Do we have spare and redundant power supplies on critical IT equipment?
- Are we adequately protected against spammers, hackers, and other attacks?

- Is our senior management committed to spending enough on disaster recovery?
- Are our networks hacker-proof?
- Do we have proper anti-virus protection?
- Are all equipment, tapes, etc, labelled properly?
- Do we have fire-proof safes to store backup tapes and important documents?
- Do we have an offsite to store important documents and tapes?
- Is there a proper change management board to approve all technical changes to the infrastructure?
- Are our telephones supplied by at least two or more different service providers?
- Is our office public address system audible in every nook and cranny?
- Are there enough static eliminators in fire hazard areas, data centres, etc?
- Are our electrical system and wiring of the proper standard?
- Do we have proper UPS and electric generators to handle long power outages?
- Are we backing up all important data every day?

Questions on health and safety
(Yes, No or N/A for each)

- Do all staff know how to operate the fire extinguishers?

- Does everyone know where the fire exits are, and are they marked clearly?

- Is the company water supply free from pollution and safe for drinking?

- Are there any harmful and hazardous materials stored in common areas?

- Do we have enough emergency lamps, torches, etc, at necessary places?

- Is our fire alarm system in working condition?

- Do we have 24x7 surveillance in critical areas?

- Is there any water seepage in critical areas?

- Do fire safety experts inspect our building periodically?

- Do we have emergency medicines and first aid facilities on the premises?

- Do we periodically have a clean-up drive to eliminate all hazardous and inflammable material around the office and premises?

- Do employees smoke cigarettes inside the office?

- Are our smoke detectors in working condition?

- Are the paths to the fire exits free from unwanted materials, boxes, etc?

- Do we have periodic and surprise fire drill and building evacuation exercises?

- Do we have posters, e-mails, newsletters, etc, that can be used to create awareness among employees?

- Are all emergency numbers readily available?

- Is our cafeteria clean and hygienic?

Questions on financial and legal issues
(Yes, No or N/A for each)

- Do we have sufficient insurance coverage for all critical equipment?

- Are we sure that all critical equipment is covered under vendor maintenance agreements?

- Do we have sufficient capacity to meet reasonable or sudden high demands?

- Do we have backup copies or scanned copies of every important document?

- Are our important paper documents safe?

- Are we storing all financial information in a highly secure location?

- Are we following professional management practices to avoid employee harassment, litigation, workplace bullying, legal complications, etc?

- Are we following all government and local tax laws ?

- Are adequate budgets available to cover disaster recovery, business continuity, etc?

Questions on people
(Yes, No or N/A for each)

- Do we have a list of all current emergency contact numbers?

15: Plenty of Questions

- Do we have enough technical staff to handle major emergencies and disasters?

- Are there any critical business or technical functions that are handled/known only by one person? Is there somebody else who can handle those functions if the primary person falls sick or quits or dies?

- Have employees installed electrical appliances such as coffeepots, radios, mobile chargers, etc? These appliances can cause electrical fires by short circuits. Building maintenance staff must be able to educate staff regarding the problems these appliances can cause.

- Do any of our IT staff consume excessive alcohol or take drugs?

- Are we paying best industry standard salaries to critical staff so that we don't have frequent resignations?

- Have we tried contacting all our key and critical staff after office hours or on weekends just for a mock exercise?

- Is there a crisis management team to handle crisis situations?

- Do we have a user awareness programme or training for employees on disaster recovery and preparedness?

- Is there a specialized and dedicated department to handle disaster recovery, business continuity, crisis management, etc?

CHAPTER 16: HOW DO I GET STARTED?

'The journey of a thousand miles must begin with a single step.'

Lao Tzu

Disaster recovery and business continuity is complex and involves much cost and efforts. In order to get started it is necessary to first have a plan and an initial scope. A plan need not or cannot be accurate or detailed from day one. It evolves and matures over time depending on learning exercises, roadblocks, mistakes, etc. Most business managers think that business continuity is primarily an IT department's job. It is not. Though IT is used extensively in businesses it is not the responsibility of the IT department alone, nor can they be blamed for business losses if some critical IT systems fail. Many departments and angles must be involved, many angles explored, and budgets prepared.

Several types of plan are required by an organization to prepare for various emergencies. These plans should cover contingencies for an organization's IT systems, business processes, facilities, people, etc. For example, if all key programmers in an important critical project quit suddenly due to some reasons like getting snatched by a competitor or not liking a bad boss then the organization may grind to a halt even though the IT and other systems may be in good condition. Hence a people-related contingency plan like having exit agreements, additional programmers, etc, can be prepared to take care of such situations.

It is possible to get or buy ready-made detailed templates and procedures for business continuity, IT contingency, people contingency, etc. Many of the detailed templates and procedures assume or recommend that every organization can afford to have fully-fledged dedicated teams – a situation management team, incident management team, crisis management team, response team, damage assessment team, recovery team, etc. Such diverse departments may be possible or necessary in big organizations, but it is rarely possible in small and medium-sized organizations to have dedicated manpower for such functions. Often it becomes the responsibility of a single person or a small team to take on added responsibilities of disaster recovery, business continuity and so on.

Simplicity and practicality become essential and hence it is always recommended to have simple, jargon-free, practical plans and checklists.

This chapter does not delve too much into academic or theoretical definitions, and does not assume a multiplicity of such teams, as this book is primarily for small and medium-sized organizations.

Some of the common plans that an organization should have are listed below.

Common types of plan

- **Business continuity plan:** This will normally provide procedures for sustaining essential business functions, people engagement, crisis management, etc, during a lengthy disruption. This plan can also have extensive dependence on IT related processes, as nowadays business processes are heavily dependent on IT systems.

- **DR or IT contingency plans:** These are mainly technology related and provide procedures to handle IT related disasters. For example, how does the IT department recover data from an important server if there is a major hardware fault? Depending on the nature and duration of an IT fault it may be necessary for an organization to invoke business continuity. For example: a bank's main computer may fail during working hours and not be expected to be rectified for several hours. This will be a case of disaster recovery. Simultaneously, the bank managers allowing customers to withdraw and deposit cash based on paper forms and signatures is a form of business continuity.

- **Crisis communication plan:** This will mainly cover procedures and rules for providing information to media, press, government, etc. Corporate communications is a very sensitive matter and must be handled by responsible and knowledgeable staff. For example, if there is indeed a major IT or some other disaster in a reputable company, regular employees must not be allowed to speak or blabber anything they think to the press or a journalist. Such things can have disastrous consequences in various forms. Only responsible senior staff should provide information related to the disruption and in a proper manner to prevent panic, stock market crash, unnecessary rumours, etc.

- **Security incidents plan:** Nowadays organizations can have cyber attacks on their websites, networks and so on. For example, someone can hack into a reputable company's website and deface it. So it is necessary to have a proper plan to handle such incidents. For example, the plan can include a procedure to give a proper press release to the media during such incidents.

- **People-related contingency plan:** This can cover processes and procedures to handle people-related disasters like accident, death or exit of key staff, epidemics, etc. It can even cover aspects like kidnap and murder of senior and other staff in certain types of organizations depending on the area in which they may be working.

- **Other plans:** Depending on the nature of an organization.

Starting a DR or BC programme in an organization

In the last few chapters we have discussed concepts like the importance of disaster recovery, business continuity, various IT and non-IT issues, etc. But the important question is 'how and where does an organization start to implement an alternative DR setup? '

This sample checklist provides a series of steps that can be followed by an organization intending to establish a DR or BC programme. Some of the steps can be done in parallel.

Step 1: Approvals and paperwork

Before any work can be initiated it is necessary to get the required senior management commitment, cost approvals and paperwork in place.

- Prepare and get an approval for a detailed DR or BC proposal for the organization covering every important or critical business function. This will normally be the responsibility of the CTO, CIO and senior business managers. The components of such a proposal were explained earlier in the question on getting senior management commitment.

- Discuss in detail with senior management, relevant department heads, IT support staff, etc, to get the entire proposal approved, along with the necessary budgets.

- Clearly agree upon what will be provided, and what will not, in writing. This is very important to avoid expectation troubles later.

Step 2: Identifying internal manpower

A DR or BC setup cannot be established and maintained by a single person, though a single person in small organizations may oversee it. It involves a lot of teamwork and co-ordination from several areas.

- Identify qualified and trained internal staff who will be responsible for various DR or BC activities.

- A DR or BC committee can be formed involving technical and non-technical staff headed by a DR manager.

- Clearly identify the roles and responsibilities of each of the staff. Also identify alternative staff members for each should any of the DR or BC members be unreachable or travelling, etc.

- Prepare detailed internal DR documentation for each of the critical business functions identified for DR. The documentation must contain what, how, where, when, etc. For example, if the payroll server is a DR or BC item, the documentation must contain complete details for installing, maintaining and synchronizing data into an alternative server housed in the DR or BC site. If there are ten other servers identified for DR, then each server must have its own specific and detailed documentation.

Step 3: Identify external manpower

It is important and necessary to involve external manpower like vendors, consultants, etc, to assist an organization in their DR activities. External assistance will be very important, perhaps critical. For example, if a vendor is not able to provide critical spares and technical assistance during a disaster, the internal staff may not be able to proceed further.

- Identify appropriate vendors for each of the critical systems and services.

- Establish clear service level agreements and commitments with penalties if necessary for non-compliance.

Step 4: Identify the alternative site

- Identify a suitable alternative site based on the various factors considered as 'musts' and 'wants'. Musts are mandatory like electrical power, telephones, communication facilities, computers, 24x7 access, etc. Wants will be nice-to-have things, but not mandatory, eg, things like air conditioning, pantry facilities, etc.

- Spruce up the alternative site with all necessary facilities: electrical power, seating arrangements, telephones, UPS, data centre, drinking water, security, storage, parking, etc.

Step 5: Get equipment

- Buy, lease or rent all the necessary IT and non-IT equipment with sufficient capacity and horsepower for DR and BC activities.

Step 6: Install and test equipment

- Get the equipment and software installed by appropriate vendors.

- Configure the equipment to meet your business requirements. For example, suppose your critical data is inside a database in the main site. Then, the alternative site database server must be configured to be identical to the main site in all respects to handle data synchronization.

- Label all equipment clearly.

- Test the equipment with dummy data, dry runs, etc.

- Have manuals, documents, checklists, procedures, etc, handy. A detailed checklist for each business function test must be available. For example, the finance department must have a checklist of the things they need to test in the DR site during the mock run, and in the event of a real disaster. This pre-defined checklist is necessary to ensure that the alternative systems provided will perform their business functions.

Step 7: Maintain the DR readiness

Establishing a DR or BC site is not a one-off task. It must be maintained to readiness at all times. Once the site has been established it is necessary for the identified team to periodically or continuously maintain the site. This includes:

- Establish data synchronization, either automatically or manually.

- Keep copies of all necessary software and documentation.

- Keep copies of every important documentation.

- Conduct dry runs or mock exercises periodically. Close all gaps and deficiencies noticed during dry runs.

- Keep the servers and other equipment updated and maintained with latest patches, anti-virus, hardware maintenance, etc.

- Upgrade equipment and services as necessary.

- Decommission unwanted equipment and services or add new services depending on business needs and growth.

- Keep the BCP and other documents updated regularly, and whenever there are additions or modifications to critical services.

Step 8: Get an external opinion and audit done

If your systems and policies permit, get an external consultant to conduct an independent audit of your DR or BC systems. Sometimes the most trivial issues could have been missed out by oversight, which an independent third party may identify. Alternatively, the consultant could offer a better, easier or more cost-effective way of doing many things. They could share some of the best practices followed in other reputable organizations.

Step 9: Tell everyone

All employees of the organization must be periodically educated about DR and BC. They need to know what the organization has planned in the event of a disaster happening and what they, the employees, will or will not be expected to do in such circumstances. User education can be done by using slide presentations, video clips, third party educational services, bulletin boards, e-mail, etc. The periodicity of the information is important. It cannot be done just once and expect that everyone has understood everything in one go. User education, training, etc, is a continuous and essential DR/BC process.

Preparing a business continuity plan

The primary objectives of any business continuity plan will be to:

- Ensure the safety of all employees and other personnel of an organization.

- Have minimum customer or reputation impact, and be in a position to keep essential internal and external business functions alive.

- Ensure that the organization is able to restore normal business functions as soon as possible.

- Meet audit and regulatory requirements, mandatory or otherwise.

The essence of business continuity lies in the following thinking:

Our organization uses 'something' that is essential for our business and revenue. If that 'something' fails suddenly what

For example, that 'something' can be as basic as a telephone or a single computer used by a lone businessman, or as complex as a series of massive mainframes used by a large corporation.

A business continuity plan starts and revolves around the following thinking:

- What are the things and systems that we consider critical for running our business?

- What are the various ways in which these critical systems and functions can stop?

- What do we do should such a situation arise?

- How long can we tolerate the disruption? What is in our control, and what is out of our control ?

- Who will be responsible to ensure that such a situation does not arise, or if it does arise how do we tackle it with a minimum of financial and other losses?

- If such a situation does occur what are the alternatives we can use to continue business?

The thinking and documentation can get more granular by asking more questions about each one of the critical systems to frame a BC plan with individual contingency plans. A BC Plan gives an overall picture for an organization, whereas a contingency plan gives specific recovery details for each of the critical systems mentioned in a BC plan.

Having a workable BC plan is critical for any organization. It determines and documents the necessary processes and procedures that will be initiated if normal business is interrupted beyond acceptable timeframes. The interruption could be due to a technology failure, natural disaster, fires, civil disturbances, etc. However, business continuity need not always be a technical solution, and can even be a simple manual solutions if possible or acceptable.

In large organizations it is the responsibility of each business or department head to ensure that they have a detailed BC plan for their respective unit. In smaller organizations business continuity can be on a smaller scale and handled by a limited number of responsible staff.

It should be noted that each plan is unique to every business and cannot be generalized into a 'one size fits all' approach. It also depends on the industry type. Secondly, it is always preferable to have one single document as an organization's BC plan. Multiple documents can cause maintenance problems. It is possible in small organizations to have a single BCP document, though the inputs may come from many departments. In large organizations it may not be possible to have a single document outlining every business function. In such cases it is better to split the organization into logical functions and delegate the BC responsibilities to individual groups to manage. A sample BC plan with contingency plans for each critical system for a hypothetical organization is shown below.

Sample BC plan for a hypothetical company

Document date: 15 January 2007

Introduction and purpose: Organizations today are subject to numerous IT and non-IT threats and disruptions. The type of disaster can range from small to severe scale affecting all business units and critical functions. Hence effective planning and preparation is required to ensure that our organization is shielded from such threats and risks. While it is not possible to safeguard from all types of threat and risk, it is within our control and financial ability to minimize most of the major risks that may threaten the systems essential for running our business and processes.

This document is intended to assist in ensuring and initiating business continuity in the event of a major disruption to normal operations in our organization. The plan outlines the roles and responsibilities of the managers and teams who will be involved to perform necessary tasks to deal with business interruptions beyond acceptable timeframes.

Document owner: Mr Charles, Business Continuity Manager. Phone: 2292343; mobile: 98762-12345.

Location of this document: The latest version will be available in the following places:

- Main site: Two hard copies in the fire-proof safe number 1 situated in the security room, ground floor. Keys available from security personnel.
- DR-BC site: Two hard copies in the fire-proof safe number 1 situated in the security room at the DR site. Keys available from security personnel.
- One soft and one hard copy with each of the business heads.
- Online copy on the company Intranet.

Periodicity of update: Document to be updated every three months, in the last week of the quarter end.

List of all critical assets and systems: The following is a list of all the critical IT and non-IT assets and systems essential for our business:

- All data on servers A, B, C and D, situated in the data centre.
- Communication systems E and F.

- Finance, sales and technical documentation and paperwork (electronic and hard copies).

- All legal documents.

- Key staff of departments A and B and senior management.

- Any other important equipment.

Each of the above should have an individual detailed contingency plan, which may be added as an appendix to the business continuity plan. The contingency plan will have to be prepared by experienced staff knowledgeable about that particular system.

Risks and impacts: The organization will incur the following potential or guaranteed financial and reputation losses should one or more critical systems fail during business hours:

System	Anticipated losses
A	US$5,000 per day
B	US$7,000 per day
C	Reputation losses and government penalties if not recoverable beyond 48 hours

What classifies as a disaster: The following situations are classified as disasters:

- Loss of the data centre due to fire.

- Major technical faults in critical computer systems.

- Denial of access to main building. Inaccessibility of the main site for any reason (eg, a fire in the next building).

- Loss of any documents.

- Loss of key personnel due to accidents, death or competition.

Constraints: The list of constraints and other factors that need to be considered, eg, staff shortage, travel delays, etc.

Limitations: List all possible limitations of the DR site, costs, logistics issues, etc. For example, limited number of telephone lines.

Risks: A list of risks to be considered. For example, a virus attack in the main site could also affect the DR site.

Location of the DR or BC site: Address, phone numbers and directions.

What is available in the DR site: List all equipment, facilities, etc, available.

What is not available: List all equipment, facilities, etc, not available.

Who can initiate a DR or BC: The following persons will be responsible for initiating or declaring a DR or BC situation.

- Managing director

- CTO and CIO

- Administration manager

Only authorized and responsible persons should have this authority.

Scope of this document: The document covers DR and BC planning for the following critical business functions and areas:

- Data centre disasters at main site

- Finance applications on the following systems (list)

- Sales processing system
- Document losses in main site
- Exit of key staff

Out of scope: This document does not cover the following scenarios:

- Loss of both main and DR-BC site for any reason.
- Earthquakes, terrorism, sabotage, civil disturbances, acts of war and other situations beyond the organization's control.

Acceptable outages: The following table outlines the acceptable outages for each business critical system:

Sl No	Critical system	Function	Owner	Agreed RPO	Agreed RTO
1	A	X	HR		
2	B	Y	Finance		
3	C	Z	Sales		
4	D	T	CTO		

For example, a key sales processing system may have an acceptable outage of six hours, so a power breakdown expected to be restored back in a couple of hours is not classified as a disaster.

16: How do I Get Started?

Assumptions: The following assumptions have been made when developing this plan:

- All stakeholders have already agreed to the minimum acceptable level of service, RTO and RPO for each of the critical business functions.
- All necessary data and documents are held at the DR-BC site.
- The DR-BC site is always in a state of readiness.
- Periodic rehearsals have been conducted and gaps closed.
- The business heads and managers have personally verified and agreed that the rehearsals conducted are acceptable for their respective business functions.
- Vendors are able to fulfil commitments during a disaster as per their service level agreements.
- Senior management will provide necessary financial assistance during a disaster or crisis.

Who does what and how: In a disaster scenario who will do what, what sort of approvals (verbal or written) are necessary, etc.

Emergency team and key personnel: The following staff and departments will be responsible for handling and managing identified disasters in our organization:

- Senior management team: the senior management team consisting of the CEO and the following persons: Mr X, Mrs Y and Mr Z.
- Crisis management team, consisting of two senior managers.
- IT team: the IT department headed by the CTO.
- Finance team: the finance department headed by the CFO.
- Office admin team: the office administration department headed by the admin manager.

- Identified business unit representatives.
- External vendors and suppliers.

Meetings: The emergency team and key personnel will meet every three months to discuss, approve and finalize all matters, upgrades and new issues related to business continuity. The business continuity manager will be primarily responsible for ensuring that all areas are covered and will chair the meetings.

Responsibilities: The responsibilities of each of the above teams are as follows:

Senior management	Responsible for evaluating seriousness of the disaster or crisis. Declare or invoke the business continuity plan. Provide necessary financial support. Determine recovery priorities and resource assignment. Communication to customers, stakeholders, board of directors, press, etc.
Crisis management team	Crisis communication: responsible for ensuring that the situation is under control from unnecessary panic. Provide press releases, communicating with the media, etc. Post-recovery communications.

IT team	Provide technical assistance for restoring data.
	Establish agreed communication facilities.
	Provide alternative or agreed technical work-arounds wherever possible.
	Provide technical support and guidance.
Admin team	Provide necessary assistance related to, material and staff movement, transportation, security arrangements, etc.
	Logistical support.
Finance team	Provide necessary and emergency financial support for activities related to the BC plan like emergency purchases, hiring of equipment, travel costs, etc.
Business unit reps	Responsible for having trained staff and necessary documentation to conduct essential operations.
	Run respective operations from alternative site.
External vendors	Provide necessary technical assistance, including spares, on-site support staff, telephonic support, etc, as agreed upon in their respective SLAs.

BC management team and structure: An updated list, contact numbers and organizational structure of all members responsible

16: How do I Get Started?

for DR and BC plan activities. Also list alternative backup members for each person.

Company contacts: Contact list to be reviewed for update every three months and immediately on any known changes.

Primary	Contact details	Alternative	Contact details
Mr A	Address 1 and phone number	Mr B	Address 2 and phone number
Mr C	Address 3 and phone number	Mr D	Address 4 and phone number

Vendor or supplier contacts: Contact list to be updated every two months:

Company	Address	Contacts
Sirius Computers		
Dell Computers		
Sun Cabling Corp		

Critical document locations: The following table contains the document locations of all critical business functions:

Documents of	Location
Finance department	Fire safe 1
Sales department	Fire safe 2
Technical manuals	Fire safe 3
Passwords	Fire safe 4

DR-BC plan scenarios: The document covers a high level DR-BC plan for the following scenarios (detailed contingency plans for each are available in the appendix of the plan):

Scenario 1: Loss of finance applications	
Probability	Medium
Business owner	Mr A, finance manager
Impact	High
Possible causes	Virus, disk crash, power failure
Functions affected	Payroll, payments

RTO and RPO	One business day
Recovery	Expected beyond one day
Action	Start manual and paper-based processes if possible. If not, move finance staff to BC site to use alternative systems. Inform all senior managers about what has happened and when it is likely to recover. Call vendor X.
Responsibilities	Finance manager
Mitigation, alternatives	Department to use standby finance system in BC site. Paper entries also to be done for post-recovery operations.
Post-recovery	Differential data to be uploaded to main system

Scenario 2: Loss of key project staff	
Probability	Medium
Business owner	Mr X, project head
Impact	High
Possible causes	Competition, accident, resignations
Functions affected	Project A
Time to get equivalent staff	One month
Recovery	
Action	Lease other department staff part-time. HR to hire additional programmers from ABC consultancy immediately. Ensure full and proper written handover to person or persons taking over. Inform customers about possible delays in shipment and other issues.

Responsibilities	Respective project managers
Mitigation and alternatives	Have additional staff to handle such emergencies
Post-recovery	

Scenario 3: Loss of data centre

Probability	Medium
Business owner	Mr A, finance manager
Impact	High
Possible causes	Fire, major power failure, LAN failure
Functions affected	All
RTO and RPO	One business day
Recovery	Expected beyond one day
Action	Move all critical departments to BC site. Inform all senior managers about what

	has happened and when it is likely to recover. Inform customers about possible delays in shipment and other issues. Call Vendors X, Y and Z.
Responsibilities	CTO
Mitigation and alternatives	All business critical functions to work from BC site until data centre is rectified. Paper entries also to be done for post-recovery operations.
Post-recovery	Differential data to be uploaded to main system.

Similar tables with additional information can be prepared for each of the critical business functions for DR-BC. Each of the above scenarios should have a contingency plan with detailed procedures, plans, etc.

Staff evacuation: Staff are expected to evacuate the building if the main siren is switched on.

In the event that it becomes essential to evacuate all employees due to a fire, bomb threat or some other emergency it is essential to have a proper evacuation procedure clearly documented and made available to all employees. Each employee must be trained on the evacuation procedure by making him or her aware of all the fire and emergency exits, assembly points, safety precautions, etc.

16: How do I Get Started?

Assembly point: The following is the assembly point for employees after evacuation till further instructions are provided:

• Football ground near station: address.

This location has been identified as the safe assembly point for all employees in the event of any evacuation requirements.

A safe assembly point within reasonable walking distance from the organization's main building should be identified. Each employee has been provided with a booklet outlining the evacuation procedure, assembly location, do's and don'ts. An evacuation rehearsal has to be conducted by the Business Continuity Team every quarter.

Crisis communication: to be done by the Crisis Management Team, headed by Mr Thomas.

Coverage: Who needs to be informed about what? Who will inform the customers, the press, and other stakeholders? This is to prevent rumours and other inappropriate information from leaking out that can have serious reputation repercussions and exacerbate the crisis. For example, a journalist can blow a minor issue out of proportion causing serious public relation issues.

Awareness training: An awareness training regarding disaster recovery and business continuity to be conducted every quarter. Training will be conducted by the training department.

Dry run schedule: All departments must conduct a mock rehearsal of their respective business functions at the BC site every quarter. The results and issues must be submitted to the business continuity manager for resolution or work-around.

Reports: Necessary reports and action items after each mock run.

Restoration phase: After the recovery of the main systems it will be necessary to terminate the business continuity activities and resume normal functions. This will involve the reverse of various activities done during initiation of the BCP. Some of the activities will be as follows:

- Stopping the activities in the DR-BC site.
- Relocation to main site.
- Restoration or re-entry of differential data into the main site systems.
- Informing customers, stakeholders, press, etc.
- Informing all relevant departments and staff.
- Unwinding activities.
- Resumption of normal functions at main site.
- Learning exercises.

Preparing an IT contingency plan

A BC plan will mainly provide a high level overview to senior management about an organization's overall business continuity capability, or lack of it. It will not directly contain detailed technical stuff that can be understood only by IT staff. For this we will need an IT contingency plan to provide the IT nuts and bolts details to recover each critical IT system. An IT contingency plan is usually very specific and goes into granular details that only department specialists can understand and invoke. However, an IT contingency plan will be an important part of a BC plan.

The BC plan can also contain other plans, like people and non-IT contingency plans. All the contingency plans (IT, non-IT, people, etc) can become important components of the overall BC plan. For example, if there is a critical finance server running an important application, a proper IT contingency plan is needed to cover emergencies like disk crashes, virus attacks, data corruption, power supply failures, etc, for only that specific application. Similarly, a people-related contingency plan could have specifics on how to handle people-related issues like sudden death, resignations or accidents to key staff.

Sample IT contingency plan for a critical server

Purpose: This is a contingency plan for our critical finance system running the Star Application used by all finance, sales and senior management. Interruption to this system beyond acceptable limits during business hours can result in significant financial and other losses to the organization. This document is part of the overall business continuity plan for the organization.

SYSTEM NAME: FS-1	
A. GENERAL DETAILS	
Asset number	45
Computer name	FS-1
Operating system	Windows 2000
Location	Data centre – main rack
Insurance details	ABCD

Hardware support vendor details	Address and phone
Hardware support contract number	36
Contract validity	Jan 07 to Dec 07
Spares available on site	One disk and one power supply inside the spares cabinet
DR machine available?	Available in readiness in DR site
Data backup	Every night at 11 pm on attached tape drive
Data sync to DR server	Every morning at 2 am, automatic. Complete data is synced.
Software media and manuals	Available in data centre and duplicate in DR site
Critical application	Star application Version 5
Software vendor	Star Systems, address, contact
System owner	Finance department

16: How do I Get Started?

Importance	High
Recovery priority	High
Possible failures	Disk crash Data corruption File deletion Power supply failures Entire system loss
Areas affected by this system failure	Sales department Finance department Order processing
Approximate number of users affected	100+
Server maintained by vendor	Star Computers, address, contact
Maintenance details	9 am to 6 pm Monday to Friday only
Vendor coverage	All hardware

16: How do I Get Started?

All IT problems to be notified to	Finance department; phone: 777
Other info 1	
Other info 2	

B. IT CONTINGENCIES	
TYPE	**PLAN**
Disk crash	Re-activate disk if possible. If not possible, call vendor support on 2345678 to replace failed disk or other component with disk of 36 gigabytes part number-2345. Mention support contract number 36 if asked by vendor. Restore data from previous day's tape. Restore complete data from session-1 from tape. How to restore: refer to RESTORE-FS1.DOC.
Power failure	Call vendor to replace power supply. Verify whether data is intact.

	Restore previous day's data if corrupt. How to restore: refer to RESTORE-FS1.DOC.
Major problem: server not recoverable for > 48 hours	Inform Business Continuity Manager Staff to use alternative server BC-FS1 in DR-BCP site until main system can be rectified or recovered. Differential data to be keyed in manually after main system is restored. Call vendor and inform of situation and to seek assistance. If main server has been totally destroyed inform finance department to initiate insurance claims for Asset Number 45. Continue further operations from BC site or transfer system to main site as appropriate.

Similar plans with more granular details as applicable can be prepared for each of the critical IT systems owned by the organization.

Conducting a mock run

After having established a DR or BC site it is absolutely necessary to keep it in readiness at all times. It is not a one-off exercise done to suit an audit, or simply to please somebody. Conducting mock runs can give valuable insights and bring out various surprises and serious gaps. It is necessary to conduct periodic mock runs involving all relevant staff to prove or disprove that the alternative site is capable of handling disaster scenarios. Some of the important points to be noted are:

- Planned mock runs must be conducted periodically, say once every three months.

- Surprise mock runs can also be conducted to find out deficiencies.

- Surprises and deficiencies can be observed.

- All gaps and deficiencies noted must be corrected as soon as possible.

- End-users and relevant staff will then be aware of what to do in a disaster situation.

- Senior management and stakeholders can be convinced of business continuity in the event of disasters.

- You can check the commitments of vendors to see if they can actually walk the talk.

Steps to conduct a mock run

- Plan a convenient date, time and duration for a mock run.

- Inform all members of the BCM team.

- Inform all relevant staff and end-user departments who will be necessary for conducting mock tests on their respective systems. For example, the finance department staff will be required for testing the payroll systems.

- Get attendance commitment from all concerned.

- Arrange transportation and other facilities.

- Technically or logically disconnect the main systems from the DR site. The assumption is that the main site has been hit with a disaster and all concerned departments have to conduct essential business operations from the alternative site.

- Inform the relevant staff to begin their respective tests from the pre-defined checklists they have been provided with. For example payroll checklist, sales checklist, etc.

- Inform the relevant staff to note down all issues, major and minor, in a clear document, form or template.

- After the mock run is over study the issues and problems experienced by each department or business unit.

- Make a list of top priority ones and arrange to resolve them as soon as possible.

- Conduct another mock run after resolving major issues.

- Get a sign-off on things that met requirements.

- Ongoing, the cycle and mock runs can go on periodically until the alternative site can become a fully tested and proven DR-BC site.

APPENDIX 1: SOURCES OF FURTHER INFORMATION

Websites and publications

Several excellent websites and publications are available for those who want to go into the depths of disaster recovery. Visit the following websites related to DR and BC:

www.itgovernance.co.uk/page.business_continuity: Leading UK provider of information and advice in the rapidly-expanding field of IT governance, regulatory compliance and information security. Their website is the best one-stop-shop for comprehensive corporate and IT governance information, advice, guidance, books, tools, training and consultancy. Their BC templates are particularly useful – a single template for smaller organizations and a set of templates for larger ones.

www.rothstein.com: An excellent website that offers a complete range of DR services. Contains a wealth of information related to DR and BC, with plenty of white papers. Rothstein Associates was established in 1985 as a management consultancy. Since 1989 they have marketed over 1,000 books, software tools, videos and research reports. Since 1994, they have themselves published over 70 books and software tools. Rothstein agencies also offer consultancy, newsletters, workshops, etc, related to DR and BC. Rothstein consultancy services can work with your team to craft a custom strategy and a tactical plan tailored to your culture, environment, style and budget.

www.itilsurvival.com: ITIL is a set of IT best practices that organizations are encouraged to implement to ensure that their IT infrastructure is supported and managed in the best possible way. Implementing ITIL is one of the best prescriptions for avoiding and managing IT-related disasters in organizations. ITIL has now become the de facto standard in delivering IT services for all types of organization. This site offers a range of topics on ITIL that can either be used immediately or be customized to your organization's requirements.

www.tekcentral.com: An information systems technology directory. Contains hundreds and hundreds of useful articles organized by topic and area. This site is a cool study guide on various modern IT topics.

www.drj.com: This is a journal dedicated to BC. Offers online and print edition subscription for the *Disaster Recovery Journal.* In addition videos, CD-Roms, articles, etc, are also available on the site. The magazine has thousands of subscribers worldwide. DRJ also sponsors various conferences and seminars related to disaster recovery.

www.bitpipe.com: An excellent website for in-depth information technology content, white papers, product literature, web casts, analyst reports and case studies. Requires one to register to access various useful documents.

www.disasterrecoveryworld.com: A neat website that catalogues various products and services related to DRP like business impact analysis, risk analysis, software tools, etc.

www.drii.org: An organization involved in training and certifying professionals in DR management. DRI International (DRII) was first formed in 1988 as the Disaster Recovery Institute in St Louis by a group from industry and from Washington University in St. Louis who forecast the need for comprehensive education in business continuity. The goals of the organization are to:

- Promote a base of common knowledge for the DR/BC industry through education, assistance and publication of the standard resource base.

- Certify qualified individuals in the discipline.

- Promote the credibility and professionalism of certified individuals.

DRII sets standards that provide the minimum acceptable level of measurable knowledge, thus providing a baseline for levels of knowledge and capabilities. Offers various courses, basic to advanced.

DR and BC consultants

Paul Kirvan

Paul F Kirvan, is Editor-in-Chief of *Contingency Planning and Management* (CPM) magazine, responsible for editorial content and direction for the magazine. He offers consultancy on DR and BC planning for LANs, WANs and telecommunications, project management, documentation, training, etc.

His professional qualifications are: Fellow, Business Continuity Institute (FBCI); Certified Business Continuity Professional (CBCP); Certified Information Systems Security Professional (CISSP) and NARTE Class 1 Certified Telecommunications Engineer (NCE).

More details are available at: *www.contingencyplanning.com* and *www.disaster-resource.com/articles/wwkirvan.shtml.*

Philip Jan Rothstein (FBCI)

Philip Jan Rothstein is a management consultant experienced in a broad variety of strategic roles. As President of Rothstein Associates Inc since 1985, his mission has been management of profound business risk in the face of unpredictable circumstances. His consulting emphasis is on business continuity, disaster avoidance and recovery, and continuous availability of high technology business environments. He was elected Fellow by the Business Continuity Institute (BCI) in 1994 in recognition of his substantial contributions to the business continuity and disaster recovery industry.

He is the publisher of *The Rothstein Catalogue on Disaster Recovery*, the industry's principal source for 1,000+ books, research reports, software tools and videos since 1989. He is also the editor and principal author of the book *Disaster Recovery Testing: Exercising Your Contingency Plan* (1994). He has edited or written over 40 books.

Mr Rothstein is a frequent speaker on business continuity, disaster avoidance and disaster recovery for top corporate management and at industry conferences.

Professional qualifications: FBCI. More details are available at: *www.rothstein.com/bio.html*.

Other consultants

Several other consultants who operate independently or as a firm offering various services on DR and BC can be located by using Internet search engines and the following websites:

- *www.disaster-recovery-guide.com/consultants.htm*
- *www.disaster-recovery-directory.com/consultants.htm*

APPENDIX 2: USEFUL TEMPLATES AND CHECKLISTS

'The palest ink is better than the most retentive memory.'

Old Chinese proverb

Here are a few useful and simple templates, checklists and tables that can be used to kick start disaster recovery in your organization. Note that these templates are not exhaustive or highly detailed. However, they are enough to start the data collection. Add additional sections as appropriate to your particular organization.

All these checklists can become components of an overall BC plan.

Important vendors list

It is necessary for organizations to have an up-to-date vendor list with phone numbers, contact names, e-mail IDs, mobiles, etc.

Vendor	Area	Contact
Vendor 1	All IBM servers	

Vendor 2	Oracle applications	
Vendor 3	Desktop support	
Vendor 4	Communication support	

Vendor selection checklist

Selecting the right vendor for support is of the utmost importance to any organization. If the vendor is unable to deliver as expected, organizations can get into serious business troubles.

- Is the vendor an authorized dealer of the product?

- How far is the vendor office from your organization?

- Does the vendor have adequate support staff who are trained and certified in the products they will support?

- Does the vendor offer 24x7 support if required?

- Does the vendor stock essential spares in their local office or will they need to wait for sourcing from head office?

- Can you contact the vendor through e-mail, mobile, pager, fax and web?

- Will the vendor sign a customized service level agreement and provide status reports?

- Can the vendor provide references?

- Will the vendor sign a non-disclosure agreement?

Appendix 2: Useful Templates and Checklists

DR staff checklist

It is necessary to have an up-to-date list of all employees who have been assigned DR responsibilities, along with their phone numbers and emergency contact information.

Name	Complete contact details
DR staff 1	
DR staff 2	
DR staff 3	
DR staff 4	
DR staff 5	

Critical systems checklist

Organizations must be able to identify all their critical systems to decide what is of high priority to the business.

Appendix 2: Useful Templates and Checklists

DRP-BCP item	Remarks
List of all critical systems that require DR and BC	<ul><li>Payroll system</li><li>Billing system</li><li>Sales system</li></ul>
Business priority	Very high
DR-BCP location	Address
Available DR systems in DR-BCP location	<ul><li>Computer 1: Payroll</li><li>Computer 2: Billing</li><li>Computer 3: Sales</li></ul>
Data synchronization between main and DR site	To be done by DR team every night and success or failure recorded, verified and signed.
Test documents for payroll, billing and sales	Available in DR site
DR staff and phone numbers	Mr X Mr Y
BCM staff and phone numbers	Mr R Mr T

Appendix 2: Useful Templates and Checklists

Crisis management staff and phone numbers	Mr G Mr S
Important vendors and phone numbers	Vendor 1 Vendor 2
Accepted RTO and RPO	RTO: One business day RPO: Two business days
DR-BCP document location	Folder DRP on Server 1 CD-Rom in Firesafe 1 Printouts in DR location
DR-BCP meeting	Last Wednesday, every month
DR exercise	To be done every quarter by Finance and Sales teams
Other info 1	
Other info 2	

Important data checklist

It is necessary to identify the organization's important data and how it is protected against disasters.

Data DRP item	Remarks
List of all important data	Complete payroll database Complete billing database Complete sales database All folders on file servers 1 and 2 E-mail folder on e-mail server 1
Method of backup	Full backup every day Tape drive and image file method Separate tape for each day
Frequency	Daily overnight complete backups
Tape storage	Fire-proof safe offsite
Data restore test	A sizeable amount of data to be restored to a test location from every backup tape every month and results recorded.

Appendix 2: Useful Templates and Checklists

Image backups of important servers	Done every fortnight after business hours
Standby equipment	Hard disks and power supplies Spare server Spare tape drive

Restore test template

Just taking data backups regularly is not enough. It is necessary to periodically restore some data and verify the results. The following simple table can be used to conduct a restore test of every backup tape. The table can be used for every server that is backed up. Data from each server's backup tape can be restored to a test location and the status recorded.

Sl No	Item	Remarks
1	Server name	FS1
2	Type	Finance server
3	Data selected for restore	D:\Payroll

4	Destination for restoring	E:\TEST on server FSTEST
5	Megabytes	2 gb
6	Restore status	Successful
7	Date	18 March 2007

Communication checklist

Here is a simple checklist or template that you can use to get started and prepared for communication loss. Add or modify sections relevant to your own organization.

Communication DR item	Remarks
Type of communication link at present	Two direct leased lines, each from different service provider One VPN 50 direct telephones 100 intercom lines via a pabx

Redundant lines	Available through manually-operated ISDN dial-up for the direct leased lines. Not available for VPN, but a dial-up Internet connection can be used during VPN failures. All direct phone lines are from two different service providers. Can also use mobile phones in case of direct lines failures.

Software support checklist

The following template can be used to identify and safeguard every critical or important application that an organization uses.

Item	Remarks
Software name	Domino finance application
Application criticality	High
Vendor	Domino Systems
Description	Software is used for storing, updating and reporting all company financial

	information on sales, invoices, payroll, taxes, payments and other associated financial aspects
Software version	Version 7.0
Vendor contact	support@dominosystems.com Phone: 7778978
Departments using the application	Finance and HR
Business contact	Finance manager
Installed on	Server 1 in data centre
Daily backups	Yes
Software media and manuals	Stored in firesafe 1
Is application under vendor support?	Yes. Contract number: 32
Backup system	Available at DR site
Data synchronization	Done every night by DR team

RTO and RPO	X and Y
Other info 1	
Other info 2	

Important documents checklist

Any organization will have several important paper and electronic documents that it will be necessary to protect and safeguard against disasters. It is necessary to have an accurate list of every important document the organization needs.

Document	Importance	Location	Owner
Legal	High		
Finance	High		
HR	High		
Technical	Medium		

Non-IT checklist at DR or BC site

It is not enough just to have good IT-related adequacy at the DR or BC site. Other non-IT related stuff must also be in place and functional to ensure immediate switchover in case of emergency. Some of the common non-IT essentials are listed below. Remember that a DR or a BC site will have to be a miniature version of the main site.

Item	Remarks
Air conditioning	
Adequate seating	
UPS power	
Diesel generator	
Adequate telephones	
Storage space	
Transport arrangements	
Cafeteria and eating arrangements	

Appendix 2: Useful Templates and Checklists

Toilets	
Drinking water	
Building security	
Stationery items	

APPENDIX 3: DISASTER RECOVERY TRAINING AND CERTIFICATION

'I was thrown out of college for cheating on the metaphysics exam;
I looked into the soul of the boy sitting next to me.'   *Woody Allen*

Though your organization may have sufficient internal talent and expertise to provide in-house disaster recovery, you should still have the relevant staff trained and certified on DR. Similar to Microsoft, Novell, Cisco and other training and certification, DR training and certification is becoming popular – even mandatory – in many organizations. This training provides students with international best practices and recommendations, case studies, methods and ways to implement disaster recovery. While selecting consultants for disaster recovery, it is highly recommended to select those with proper DR certifications and relevant experience in the respective fields.

DRI International *(www.drii.org)*

DRI is a reputable organization dedicated to providing certification and education in the field of DR and BC. DRI International's certification programmes acknowledge an individual's effort to achieve a professional level of competence in the industry. This recognition, in turn, helps strengthen the credibility of the profession as a whole.

This organization offers various basic to advanced training courses and certification for professionals. See below for details but for the latest information visit their website. The course fees range from about US$1,300 to about US$3,000. The complete DRI course catalogue in PDF format can also be downloaded from the DRI site.

Associate Business Continuity Professional (ABCP)

Associate level is for individuals with less than two years of industry experience but who have the minimum knowledge in business continuity planning and have successfully passed the certification examination. The ABCP certification supports entry-level proficiency and forms the basis for moving to DRII's basic certification as a Certified Business Continuity Professional (CBCP). This first certification level also targets interested individuals who have expertise in a related field (such as marketing, or emergency response), or in fields not directly related to BC planning.

Certified Functional Continuity Professional (CFCP)

This is a new certification designed for individuals with a minimum of two years of experience, who have passed the certification exam but not at the level necessary to apply for the CBCP certification, and/or who specialize in a defined, functional area such as risk analysis, data recovery, human relations, etc. Application required and continuing education activities must be documented every two years.

Certified Business Continuity Professional CBCP

DRII's CBCP certification is reserved for individuals who have demonstrated knowledge and skill in the DR/BC industry. Regular re-certification requires an ongoing commitment to continuing education and industry activities.

Master Business Continuity Professional (MBCP)

Master level targets individuals with significant, demonstrated knowledge and skill in the industry and a minimum of five years of experience as a DR/BC planner. To qualify, you must first pass the CBCP exam with a score of 85% or higher and then successfully complete the Master Case Study Exam with a score of 75% or better.

Retired Business Continuity Professional

The 'retired' category is reserved for those individuals who have achieved the CBCP or MBCP designation, and who have maintained their certification in good standing, for a minimum of five years prior to retiring from active employment or consulting in the BC planning profession. Individuals seeking this designation must obtain prior approval from the DRII administrative office.

Business Continuity Institute *(www.thebci.org)*

The Business Continuity Institute provides recognition and professional qualifications for BC practitioners. Represented in several countries, its membership brings various benefits to professionals, including recognition, networking, education and knowledge. The site also offers newsletters, best practices and so on.

BCI has five membership grades:

Student

This is for students having an interest in the subject area taking a qualifying course or subject.

Affiliate of the Business Continuity Institute

This is for an individual expressing an interest in BC management or who is a Member of an associated Institute where both governing bodies have agreed that joint membership can be offered.

MBCI Member of the Business Continuity Institute

Three years working in the profession. Demonstrated proficiency in all 10 disciplines.

FBCI Fellow of the Business Continuity Institute

Two years as an MBCI (minimum 5 years in the profession). Demonstrated thorough proficiency in all 10 disciplines. Pass a structured 1-hour interview conducted by 3 FBCIs. Made a significant contribution to the profession.

BCI courses

Some of the courses recognized and approved by the BCI are listed below. Check out their website for latest and detailed information on costs, duration, prerequisites, etc.

Implementing and Understanding PAS 56: PAS 56, the new Publicly Available Specification on BC management, provides a framework for organizations to implement and maintain a business continuity programme.

Command Centre Operation and Crisis Leadership in Business Continuity Management: This course allows participants to gain knowledge and experience through various disaster simulations of demanding and realistic situations. This will also cover how to operate from a command centre, the issues involved, learning exercises and so on.

Fundamentals of Business Continuity: This is designed to provide the basics of business continuity, risk assessment, crisis management and so on.

Telecomm and Call Centre Continuity

Implementing a BCM infrastructure: This practical course will provide students with a comprehensive understanding of

BC management. On completion of this course students will be able to successfully implement, manage and maintain BC plans within their organization.

Effective Crisis Management: This course provides students with the key abilities required to handle major crisis situations.

Other courses and sources

The wealth of information available nowadays regarding DR and BC is simply enormous. Organizations and industries are slowly realizing the immense importance of having staff and departments to look after DR and BC.

In addition to the above two reputable and popular DR and BC certifications, various other institutes, universities, consultancies and commercial training providers are also now offering courses and certifications in DR, BC, emergency planning, etc. Many large organizations have developed in-house training and courses on DR and BCP, tailored to their own specific needs and industries.

IT Governance (www.itgovernance.co.uk)

IT Governance source, create and deliver products and services to meet the real-world, evolving IT governance needs of today's organizations, directors, managers and practitioners. This includes training and consultancy, where they are one of the leaders in the UK, offering both public and in-house training.

They are very practical and business-oriented, approaching IT governance, regulatory compliance and information security issues from a management perspective. They are committed to engaging business leaders in developing and implementing information, ICT regulatory compliance and information security strategies that enable their businesses to compete effectively in the global information economy.

Sentryx Certification (*www.sentryx.com*)

Sentryx is a DR/BC consultancy and training firm offering various training courses. It is dedicated to providing a high standard of professional business continuity training and consulting services. Sentryx offers a range of BC training products and services including on-site and off-site training, books, and computer-based training (CBT) packages. Sentryx's consulting services include BC health check, business impact analysis, and BC strategy and plan development. The courses are designed to meet the requirements for professional development, certification, and corporate BC programmes. Their BC consulting services include planning process implementation, plan development, and programme audit.

Some of the excellent courses being offered by Sentryx are:

BC-201 Business Continuity Planning: An intensive 5-day training course designed to help organizations protect their businesses.

BC-301 Business Impact Analysis: A 1-day course that teaches the steps necessary to conduct your own BIA.

BC-302 Business Continuity Plan Testing: A 1-day course that teaches the principles to validate your own BC plan.

BC-401 Managing Business Continuity Programmes: teaches how to initiate, implement, and maintain an effective BC programme within an organization.

Visit *www.sentryx.com* for detailed information on all the courses being offered and the prices for each.

Survive! The Business Continuity Group (www.survive.com)

Provides BC and DR professionals with an objective forum for the exchange of information, ideas and experiences, and access to an international network of peers. Offers a free magazine, plus conferences, meetings, workshops, seminars and other resources. Also offers information about the Business Continuity Institute, providing international recognition and professional qualifications for BC practitioners.

BCM Academy (www.bcmacademy.nl)

BCM Academy is the leading European information institute for business continuity and crisis management. It offers a full range of education, training and courses in an inspirational environment; without a surfeit of information, but rather a carefully balanced combination of theory and practical execution.

Institute for Business Continuity Training (www.ibct.com)

IBCT provides customized, in-house training and workshops for corporate clients, their business units, and BC practitioners. Each of the instructors has over 20 years of operational business and IT experience, all are experienced trainers, and all are professionally-certified BC practitioners.

Disaster Survival Planning Network (www.disaster-survival.com)

US-based DSPN helps organizations develop comprehensive enterprise-wide business continuity programmes. They are a network of nationally-certified business continuity and emergency response professionals who work one-on-one with top executives, department managers, safety committees, and BC project coordinators to plan, implement, and test BC programmes. For clients who already have BC programmes, DSPN help audit, test, and update their capabilities, thereby identifying gaps in their processes and recommending enhancements consistent with best practices in the industry.

Several other excellent DR and BC programme links are available at:

www.rothstein.com/links/rothstein_recommended29.html.

APPENDIX 4: BUSINESS CONTINUITY STANDARDS

As with any emerging science or knowledge, DR and BC processes are becoming more and more mature. For example, we have almost universal standards in financial accounting, so an accountant can move from one organization to another and still be able to practice accounting. Though organizations can frame and implement their own workable BC practices very soon they hit certain limitations and roadblocks. Hence there is a need for certain standards and universally adoptable practices that can be implemented in any industry. After the recent international terrorist attacks, mega financial scandals and electronic methods of making a company bankrupt within hours governments agencies and business owners have been pushing for establishing fool-proof BC and risk mitigation methods. A standard for BC management has been mooted for many years. Standards remove the headaches of proprietary and non-portable procedures. BC standards establish a sound basis for understanding, developing and implementing BC within an organization. Secondly, adopting industry acceptable practices can give business managers confidence that their business is in safe hands, can withstand a real disaster and also withstand the scrutiny of a comprehensive audit.

PAS56 – Guide to Business Continuity Management

This is a reference document developed by the British Standards Institute and being considered by the ISO as a global BC standard (ISO 17799 for information security).

PAS stands for Publicly Available Specification. PAS56 provides an overview of the steps and activities necessary in setting up a BC management process and makes recommendations for best practice. PAS56 ensures that an organization's mission-critical and key risk scenarios are included in the BC management processes. PAS56 can be used in all organizations regardless of size or industry.

For more information visit *www.thebci.org/pas56.htm*, but note that PAS56 was superseded as the international best practice standard for BC by the new British Standard 25999, published in November 2006 and described below.

BS25999

Many professional and competent agencies around the world are engaged in establishing standards that can be implemented in any organization. One such standard is BS25999, developed by the British Standards Institute in 2006. Though this book has been written as general-purpose advice on DR and BC for small organizations, it is worth mentioning the highlights of emerging BC standards. In a couple of years time all organizations will gradually and happily adopt industry standard BC practices as is does not make sense reinventing the wheel. As organizations go global they need to prove to each other that they have and practice certain universally acceptable standards, rather than some home-made remedies and recipes for BC. This is where industry standards will help organizations. For example, we already have several standards like ISO, CMM, ITIL and other standards for running a business. Similarly, standards like BS25999 will become a standard for protecting a business and will soon become a necessary requirement for most businesses.

Highlights of BS25999

BS25999 has been developed by the British Standards Institute and essentially establishes a code of practice for BC management.

It was released in November 2006, and is also known as BS25999-1:2006. The 1 stands for Part 1. The BS 25999 series includes two standards:

- BS 25999-1:2006 Code of Practice for BCM

- BS 25999-2:2006 A Specification for BCM

BS 25999-1:2006 establishes the necessary processes, principles and terminology of business continuity management (BCM). It provides a basis or a framework for understanding, developing and implementing BC within an organization.

Part two of the standard, BS 25999-2:2006, is expected to be released in 2007 and will specify the requirements for establishing, implementing, operating, monitoring, reviewing, maintaining and improving a documented BCM system covering an organization's overall business risks. It will also specify BC controls customized to the needs of individual organizations.

BS 25999-1:2006 replaces the PAS56 standards released earlier.

The contents of the code of practice (BS 25999-1) are:

- Scope

- Structure

- Terms

- BCM

- Overview

- The BCM system

- Understanding the organization

- BCM strategies

- Developing and implementing a BCM response

- Exercising, maintenance, audit and self-assessment of the BCM culture

- Building a BCM culture

- Types and methods of exercising BCM strategies.

The BS25999 specification will specify the requirements for implementing, operating, monitoring, reviewing, maintaining and improving a documented BCM system.

Purchasing the material

The publication can be directly purchased from www.itgovernance.co.uk/products/632. At the time of writing, the cost of the book was about £90 plus shipping charges.

APPENDIX 5: USEFUL STUFF

Organizations can educate employees regarding DR and BC practices in a number of ways. DR and BC education need not always be some boring, process-oriented work. It can be made lively and exciting. Organizations can spread the word about DR and BC more effectively, and make more of an impact, by using resources such as the following:

Posters

Organizations can, for very low cost, print colourful and eye-catching posters on DR and BC issues and paste them in all strategic locations, common and security areas. Having a DR or BC related poster is more useful and meaningful than having posters of rock stars, film stars, racecars, jokes, etc. A powerful statement or tips on a poster can become very useful or handy in times of a personal or corporate crisis.

Videos

DR and BC related videos and DVDs prepared by various industry experts could be presented to all employees, or to a section of employees. Such material can also be made available as part of an organization's library. Such videos are expensive, but they are worth it.

Contests and quizzes

Organizations can have small contests, essay competitions and quizzes via a newsletter or a company intranet to make employees more aware of DR and BC practices. Gifts and rewards can also be provided.

Reward programmes

Management can initiate reward programmes for employees who report potential problems or issues that could lead to disasters. For example, an employee who detects a particular problem that could lead to a fire hazard can be given a reward for the observation and escalation. Management must always be open to listen to bad and costly news first, instead of good and happy news. Easy ways like having suggestion boxes, registers, and e-mail IDs for reporting possible disasters can also be implemented.

APPENDIX 6: DISASTER RECOVERY GLOSSARY

Activation: The implementation of business continuity capabilities, procedures, activities and plans in response to an emergency or disaster declaration; the execution of the recovery plan.

Alert: Notification that a potential disaster situation exists or has occurred; direction for recipient to stand by for possible activation of disaster recovery plan.

Alternative site: An alternative operating location to be used by business functions when the primary facilities are inaccessible: (1) another location, computer centre or work area designated for recovery; (2) a location, other than the main facility, that can be used to conduct business functions; (3) a location, other than the normal facility, used to process data and/or conduct critical business functions in the event of a disaster. *Similar terms:* Alternative processing facility, Alternative office facility, Alternative communication facility, Backup location, Recovery site.

Alternative work area: Office recovery environment complete with necessary office infrastructure (desk, telephone, workstation, and associated hardware, communications, etc); also referred to as work space or alternative work site.

Application recovery: The component of disaster recovery that deals specifically with the restoration of business system software and data, after the processing platform has been restored or replaced. *Similar terms:* Business system recovery.

Backup generator: An independent source of power, usually fuelled by diesel or natural gas.

Business continuity planning (BCP): Process of developing advance arrangements and procedures that enable an organization to respond to an event in such a manner that critical business functions continue with planned levels of interruption or essential change. *Similar terms:* Contingency planning, Disaster recovery planning.

Business continuity programme: An ongoing programme supported and funded by executive staff to ensure business continuity requirements are assessed, resources are allocated and recovery and continuity strategies and procedures are completed and tested.

Business continuity steering committee: A committee of decision makers, business owners, technology experts and continuity professionals, tasked with making strategic recovery and continuity planning decisions for the organization.

Business impact analysis (BIA): The process of analysing all business functions and the effect that a specific disaster may have upon them. (1) Determining the type or scope of difficulty caused to an organization should a potential event identified by the risk analysis actually occur. The BIA should quantify, where possible, the loss impact from both a business interruption (number of days) and a financial standpoint. *Similar terms:* Business exposure assessment, Risk analysis

Business interruption: Any event, whether anticipated (eg, public service strike) or unanticipated (eg, blackout) which disrupts the normal course of business operations at an organization location.

Business interruption costs: The costs or lost revenue associated with an interruption in normal business operations.

Business interruption insurance: Insurance coverage for disaster related expenses that may be incurred until operations are fully recovered after a disaster.

Business recovery coordinator: An individual or group designated to coordinate or control designated recovery processes or testing. *Similar terms:* Disaster recovery coordinator.

Business recovery timeline: The chronological sequence of recovery activities, or critical path, that must be followed to resume an acceptable level of operations following a business interruption. This timeline may range from minutes to weeks, depending upon the recovery requirements and methodology.

Business resumption planning (BRP): *See:* Business continuity planning, Disaster recovery planning.

Business recovery team: A group of individuals responsible for maintaining the business recovery procedures and coordinating the recovery of business functions and processes. *Similar terms:* Disaster recovery team.

Business unit recovery: The component of disaster recovery which deals specifically with the relocation of a key function or department in the event of a disaster, including personnel, essential records, equipment supplies, work space, communication facilities, work station computer processing capability, fax, copy machines, mail services, etc. *Similar terms:* Work group recovery.

Call tree: A document that depicts graphically the calling responsibilities and the calling order used to contact management, employees, customers, vendors and other key contacts in the event of an emergency, disaster or severe outage situation.

Certified Business Continuity Professional (CBCP): The Disaster Recovery Institute International (DRI International), a not-for-profit corporation, certifies CBCPs and promotes credibility and professionalism in the business continuity industry. Also offers MBCP (Master Business Continuity Professional) and ABCP (Associate Business Continuity Professional). *See Appendix 3 for more details and website address.*

Checklist exercise: A method used to exercise a completed disaster recovery plan. This type of exercise is used to determine if the information such as phone numbers, manuals, equipment, etc, in the plan is accurate and current.

Cold site: An alternative facility that already has in place the environmental infrastructure required to recover critical business functions or information systems, but does not have any pre-installed computer hardware, telecommunications equipment, communication lines, etc. These must be provisioned at time of disaster. *Similar terms:* Shell site; Backup site; Recovery site; Alternative site.

Appendix 6: Disaster Recovery Glossary

Communications recovery: The component of disaster recovery which deals with the restoration or rerouting of an organization's telecommunication network, or its components, in the event of loss. *Similar terms:* Telecommunications recovery, Data communications recovery.

Computer recovery team: A group of individuals responsible for assessing damage to the original system, processing data in the interim, and setting up the new system.

Consortium agreement: An agreement made by a group of organizations to share processing facilities and/or office facilities, if one member of the group suffers a disaster. *Similar terms:* Reciprocal agreement.

Command centre: Facility separate from the main facility and equipped with adequate communications equipment from which initial recovery efforts are manned and media-business communications are maintained. The management team uses this facility temporarily to begin coordinating the recovery process and its use continues until the alternative sites are functional.

Contact list: A list of team members and/or key players to be contacted including their backups. The list will include the necessary contact information (eg, home phone, pager, mobile, etc) and in most cases will be considered confidential.

Contingency planning: Process of developing advance arrangements and procedures that enable an organization to respond to an event that could occur by chance or in unforeseen circumstances.

Contingency plan: A plan used by an organization or business unit to respond to a specific systems failure or disruption of operations. A contingency plan may use any number of resources including work-around procedures, an alternative work area, a reciprocal agreement or replacement resources.

Continuity of operations plan (COOP): A COOP provides guidance on system restoration for emergencies, disasters, mobilization and for maintaining a state of readiness to provide the necessary level of information processing support commensurate with the mission requirements / priorities identified by the respective functional proponent. This term is traditionally used by the US Federal Government and its supporting agencies to describe activities otherwise known as disaster recovery, business continuity, business resumption or contingency planning.

Crate and ship: A strategy for providing alternative processing capability in a disaster, via contractual arrangements with an equipment supplier, to ship replacement hardware within a specified time period. *Similar terms:* Guaranteed replacement, Drop ship, Quick ship.

Crisis: A critical event, which, if not handled in an appropriate manner, may dramatically affect an organization's profitability, reputation or ability to operate.

Crisis management: The overall coordination of an organization's response to a crisis in an effective, timely manner with the goal of avoiding or minimizing damage to the organization's profitability, reputation or ability to operate.

Crisis management team: A crisis management team will consist of key executives as well as key role players (eg, media representative, lawyer, facilities manager, disaster recovery coordinator, etc) and the appropriate business owners of critical organization functions.

Crisis simulation: The process of testing an organization's ability to respond to a crisis in a coordinated, timely and effective manner, by simulating the occurrence of a specific crisis.

Critical functions: Business activities or information that could not be interrupted or unavailable for several business days without significantly jeopardizing operation of the organization.

Critical infrastructure: Systems whose incapacity or destruction would have a debilitating impact on the economic security of an organization, community, country, etc.

Critical records: Records or documents that, if damaged or destroyed, would cause considerable inconvenience and/or require replacement or re-creation at considerable expense.

Damage assessment: The process of assessing damage, following a disaster, to computer hardware, vital records, office facilities, etc, and determining what can be salvaged or restored and what must be replaced.

Data backups: The back up of system, application, program and/or production files to media that can be stored both on and/or off-site. Data backups can be used to restore corrupted or lost data or to recover entire systems and databases in the event of a disaster. Data backups should be considered confidential and should be kept secure from physical damage and theft.

Data backup strategies: Those actions and backup processes determined by an organization to be necessary to meet its data recovery and restoration objectives. Data backup strategies will determine the timeframes, technologies, media and off-site storage of the backups, and will ensure that recovery point and time objectives can be met.

Data centre recovery: The component of disaster recovery which deals with the restoration, at an alternative location, of data centres services and computer processing capabilities. *Similar terms:* Mainframe recovery, Technology recovery.

Data recovery: The restoration of computer files from backup media to restore programs and production data to the state that existed at the time of the last safe backup.

Database replication: The partial or full duplication of data from a source database to one or more destination databases. Replication may use any of a number of methodologies including mirroring or shadowing, and may be performed synchronously, asynchronously, or point-in-time depending on the technologies used, recovery point requirements, distance and connectivity to the source database, etc. Replication can, if performed remotely, function as a backup for disasters and other major outages. *Similar terms:* File shadowing, Disk mirroring.

Disk mirroring: Disk mirroring is the duplication of data on separate disks in real time to ensure its continuous availability, currency and accuracy. Disk mirroring can function as a disaster recovery solution by performing the mirroring remotely. True mirroring will enable a zero recovery point objective. Depending on the technologies used, mirroring can be performed synchronously, asynchronously, semi-synchronously or point-in-time. *Similar terms:* File shadowing, Data replication, Journaling.

Declaration: A formal announcement by pre-authorized personnel that a disaster or severe outage is predicted or has occurred and that triggers pre-arranged mitigating actions (eg, a move to an alternative site).

Declaration fee: (1) A one-time fee, charged by an alternative facility provider, to a customer who declares a disaster. *Note:* Some recovery vendors apply the declaration fee against the first few days of recovery. (2) An initial fee or charge for implementing the terms of a recovery agreement or contract. *Similar terms:* Notification fee.

Desk check: One method of testing a specific component of a plan. Typically, the owner or author of the component reviews it for accuracy and completeness and signs off.

Disaster: A sudden, unplanned calamitous event causing great damage or loss. (1) Any event that creates an inability on an organization's part to provide critical business functions for some predetermined period of time. (2) In the business environment, any event that creates an inability on an organization's part to provide critical business functions for some predetermined period of time. (3) The period when company management decides to divert from normal production responses and exercises its disaster recovery plan. Typically signifies the beginning of a move from a primary to an alternative location. *Similar terms:* Business interruption; Outage; Catastrophe.

Disaster recovery: Activities and programmes designed to return the entity to an acceptable condition. (1) The ability to respond to an interruption in services by implementing a disaster recovery plan to restore an organization's critical business functions.

Disaster recovery or business continuity coordinator: The disaster recovery coordinator may be responsible for overall recovery of an organization or unit(s). *Similar terms:* Business recovery coordinator.

Disaster Recovery Institute International (DRI International): A not-for-profit organization that offers certification and educational offerings for business continuity professionals. *See Appendix 3 for further information and website address.*

Disaster recovery plan: The document that defines the resources, actions, tasks and data required to manage the business recovery process in the event of a business interruption. The plan is designed to assist in restoring the business process within the stated disaster recovery goals.

Disaster recovery planning: The technological aspect of business continuity planning. The advance planning and preparations that are necessary to minimize loss and ensure continuity of the critical business functions of an organization in the event of a disaster. *Similar terms:* Contingency planning; Business resumption planning; Corporate contingency planning; Business interruption planning; Disaster preparedness.

Disaster recovery software: An application program developed to assist an organization in writing a comprehensive disaster recovery plan.

Disaster / business recovery teams: A structured group of teams ready to take control of the recovery operations if a disaster should occur.

Electronic vaulting: Electronically forwarding backup data to an offsite server or storage facility. Vaulting eliminates the need for tape shipment and therefore significantly shortens the time required to move the data offsite.

Emergency: A sudden, unexpected event requiring immediate action due to potential threat to health and safety, the environment or property.

Emergency preparedness: The discipline that ensures an organization, or community's readiness to respond to an emergency in a coordinated, timely and effective manner.

Emergency procedures: A plan of action to commence immediately to prevent the loss of life and minimize injury and property damage.

Emergency operations centre (EOC): A site from which response teams/officials (municipal, county, state and federal/national) exercise direction and control in an emergency or disaster.

Environment restoration: Re-creation of the critical business operations in an alternative location, including people, equipment and communications capability.

Executive/management succession: A predetermined plan for ensuring the continuity of authority, decision-making, and communication in the event that key members of senior management suddenly become incapacitated or in the event that a crisis occurs while key members of senior management are unavailable.

Exercise: An activity that is performed for the purpose of training and conditioning team members, and improving their performance. Types of exercise include: Table top exercise, Simulation exercise, Operational exercise and Mock disaster.

File shadowing: The asynchronous duplication of the production database on separate media to ensure data availability, currency and accuracy. File shadowing can be used as a disaster recovery solution if performed remotely, to improve both the recovery time and recovery point objectives. *Similar terms:* Data replication, Journaling, Disk mirroring.

Financial impact: An operating expense that continues following an interruption or disaster, which as a result of the event cannot be offset by income and directly affects the financial position of the organization.

Forward recovery: The process of recovering a database to the point of failure by applying active journal or log data to the current backup files of the database.

Hazard or threat identification: The process of identifying situations or conditions that have the potential to cause injury to people, damage to property or damage to the environment.

High availability: Systems or applications requiring a very high level of reliability and availability. High availability systems typically operate 24x7 and usually require built-in redundancy to minimize the risk of downtime due to hardware and/or telecommunication failures.

High-risk areas: Heavily-populated areas, particularly susceptible to high-intensity earthquakes, floods, tsunamis or other disasters, for which emergency response may be necessary in the event of a disaster.

Hotsite: An alternative facility that already has in place the computer, telecommunications and environmental infrastructure required to recover critical business functions or information systems.

Human threats: Possible disruptions to operations resulting from human actions (eg, disgruntled employee, terrorism, blackmail, job actions, riots, etc).

Incident command system (ICS): Combination of facilities, equipment, personnel, procedures, and communications operating within a common organizational structure with responsibility for management of assigned resources to effectively direct and control the response to an incident. Intended to expand, as situation requires larger resources, without requiring new, reorganized command structure.

Incident manager: Commands the local EOC reporting up to senior management on the recovery progress. Has the authority to invoke the local recovery plan.

Incident response: The response of an organization to a disaster or other significant event that may significantly affect the organization, its people or its ability to function productively. An incident response may include evacuation of a facility, initiating a disaster recovery plan, performing damage assessment and any other measures necessary to bring an organization to a more stable status.

Integrated test: A test conducted on multiple components of a plan, in conjunction with each other, typically under simulated operating conditions

Interim site: A temporary location used to continue performing business functions after vacating a recovery site and before the original or new home site can be occupied.

Move to an interim site may be necessary if ongoing stay at the recovery site is not feasible for the period of time needed or if the recovery site is located far from the normal business site that was affected by the disaster. An interim site move is planned and scheduled in advance to minimize disruption of business processes; equal care must be given to transferring critical functions from the interim site back to the normal business site.

Internal hotsite: A fully equipped alternative processing site owned and operated by the organization.

Journaling: The process of logging changes or updates to a database since the last full backup. Journals can be used to recover previous versions of a file before updates were made, or to facilitate disaster recovery, if performed remotely, by applying changes to the last safe backup. *Similar terms:* File shadowing, Data replication, Disk mirroring.

LAN recovery: The component of business continuity that deals specifically with the replacement of LAN equipment and the restoration of essential data and software in the event of a disaster. Also known as client/server recovery.

Line rerouting: A short-term change in the routing of telephone traffic, which can be planned and recurring, or a reaction to an outage situation. Many regional telephone companies offer a service that allows a computer centre to quickly reroute a network of dedicated lines to a backup site.

Loss reduction: The technique of instituting mechanisms to lessen exposure to a particular risk. Loss reduction involves planning for, and reacting to, an event to limit its impact. Examples of loss reduction include sprinkler systems, insurance policies and evacuation procedures.

Lost transaction recovery: Recovery of data (paper within the work area and/or system entries) destroyed or lost at the time of the disaster or interruption. Paper documents may need to be requested or re-acquired from original sources. Data for system entries may need to be recreated or re-entered.

Mission-critical application: An application that is essential to the organization's ability to perform necessary business functions. Loss of the mission-critical application would have a negative impact on the business, as well as legal or regulatory effects.

Mobile recovery: A mobilized resource purchased or contracted for the purpose of business recovery. The mobile recovery centre might include: computers, workstations, telephone, electrical power, etc.

Mock disaster: One method of exercising teams in which participants are challenged to determine the actions they would take in the event of a specific disaster scenario. Mock disasters usually involve all, or most, of the relevant teams. Under the guidance of exercise coordinators, the teams walk through the actions they would take per their plans, or simulate performance of these actions. Teams may be at a single exercise location, or at multiple locations, with communication between teams simulating actual 'disaster mode' communications. A mock disaster will typically operate on a compressed timeframe representing many hours, or even days.

Natural threats: Events caused by nature that have the potential to affect an organization.

Network outage: An interruption in system availability resulting from a communication failure affecting a network of computer terminals, processors and/or workstations.

Off-site storage: Alternative facility, other than the primary production site, where duplicated vital records and documentation may be stored for use during disaster recovery.

Operational exercise: One method of exercising teams in which participants perform some or all of the actions they would take in the event of plan activation. Operational exercises, which may involve one or more teams, are typically performed under actual operating conditions at the designated alternative location, using the specific recovery configuration that would be available in a disaster.

Operational impact analysis: Determines the impact of the loss of an operational or technological resource. The loss of a system, network or other critical resource may affect a number of business processes.

Operational test: A test conducted on one or more components of a plan under actual operating conditions.

Plan administrator: The individual responsible for documenting recovery activities and tracking recovery progress.

Peer review: One method of testing a specific component of a plan. Typically, the component is reviewed for accuracy and completeness by personnel (other than the owner or author) with appropriate technical or business knowledge.

Plan maintenance procedures: Maintenance procedures outline the process for the review and update of business continuity plans.

Reciprocal agreement: Agreement between two organizations (or two internal business groups) with basically the same equipment/same environment that allows each one to recover at the other's site.

Recovery: (1) Process of planning for and/or implementing expanded operations to address less time-sensitive business operations immediately following an interruption or disaster. (2) The start of the actual process or function that uses the restored technology and location.

Recovery period: The time period between a disaster and a return to normal functions, during which the disaster recovery plan is employed.

Recovery services contract: A contract with an external organization guaranteeing the provision of specified equipment, facilities or services, usually within a specified time period, in the event of a business interruption. A typical contract will specify a monthly subscription fee, a declaration fee, usage costs, method and amount of testing, termination options, penalties and liabilities, etc.

Recovery strategy: An approach by an organization that will ensure its recovery and continuity in the face of a disaster or other major outage. Plans and methodologies are determined by the organization's strategy. There may be more than one methodology or solution for an organization's strategy. Examples of methodologies and solutions include contracting for hotsite or coldsite, building an internal hotsite or coldsite, identifying an alternative work area, a consortium or reciprocal agreement, contracting for mobile recovery or crate and ship, and many others.

Recovery point objective: The point in time to which systems and data must be recovered after an outage (eg, end of previous day's processing). RPOs are often used as the basis for the development of backup strategies and as a determinant of the amount of data that may need to be recreated after the systems or functions have been recovered.

Recovery time objective (RTO): The period of time within which systems, applications, or functions must be recovered after an outage (eg, one business day). RTOs are often used as the basis for the development of recovery strategies, and as a determinant as to whether or not to implement the recovery strategies during a disaster situation. *Similar term:* Maximum allowable downtime.

Response: (1) The reaction to an incident or emergency to assess the damage or impact and to ascertain the level of containment and control activity required. In addition to addressing matters of life safety and evacuation, response also addresses the policies, procedures and actions to be followed in the event of an emergency. (2) The step or stage that immediately follows a disaster event where actions begin as a result of the event having occurred. *Similar terms:* Emergency response, Disaster response, Immediate response and Damage assessment.

Restoration: Process of planning for and/or implementing procedures for the repair or relocation of the primary site and its contents, and for the restoration of normal operations at the primary site.

Resumption: (1) The process of planning for and/or implementing the restarting of defined business operations following a disaster, usually beginning with the most critical or time-sensitive functions and continuing along a planned sequence to address all identified areas required by the business. (2) The step or stage after the affected infrastructure, data, communications and environment have been successfully re-established at an alternative location.

Risk: Potential for exposure to loss. Risks, either man-made or natural, are constant. The potential is usually measured by its probability in years.

Risk assessment / analysis: Process of identifying the risks to an organization, assessing the critical functions necessary for an organization to continue business operations, defining the controls in place to reduce organization exposure and evaluating the cost for such controls. Risk analysis often involves an evaluation of the probabilities of a particular event.

Risk mitigation: Implementation of measures to deter specific threats to the continuity of business operations, and/or respond to any occurrence of such threats in a timely and appropriate manner.

Salvage and restoration: The process of reclaiming or refurbishing computer hardware, vital records, office facilities, etc, following a disaster.

Simulation exercise: One method of exercising teams in which participants perform some or all of the actions they would take in the event of plan activation. Simulation exercises, which may involve one or more teams, are performed under conditions that at least partially simulate 'disaster mode'. They may or may not be performed at the designated alternative location, and typically use only a partial recovery configuration.

Stand-alone test: A test conducted on a specific component of a plan, in isolation from other components, typically under simulated operating conditions.

Structured walkthrough: One method of testing a specific component of a plan. Typically, a team member makes a detailed presentation of the component to other team members (and possibly non-members) for their critique and evaluation.

Subscription: Contract commitment that provides an organization with the right to use a vendor recovery facility for processing capability in the event of a disaster declaration.

System downtime: A planned or unplanned interruption in system availability.

Table top exercise: One method of exercising teams in which participants review and discuss the actions they would take per their plans, but do not perform any of the actions. The exercise can be conducted with a single team, or multiple teams, typically under the guidance of exercise facilitators.

Test: An activity that is performed to evaluate the effectiveness or capabilities of a plan relative to specified objectives or measurement criteria. Types of test include: desk check, peer review, structured walkthrough, stand-alone test, integrated test and operational test.

Test plan: A document designed to periodically exercise specific action tasks and procedures to ensure viability in a real disaster or severe outage situation.

Uninterruptible power supply (UPS): A backup supply that provides continuous power to critical equipment in the event that commercial power is lost.

Vital record: A record that must be preserved and available for retrieval if needed.

Warm site: An alternative processing site which is equipped with some hardware, and communications interfaces, electrical and environmental conditioning which is only capable of providing backup after additional provisioning, software or customization is performed.

Work-around procedures: Interim procedures that may be used by a business unit to enable it to continue to perform its critical functions during temporary unavailability of specific application systems, electronic or hard copy data, voice or data communication systems, specialized equipment, office facilities, personnel, or external services. Similar term: Interim contingencies.

Printed in the United Kingdom
by Lightning Source UK Ltd.
119520UK00001B/159